Leckie
the education publisher
for Scotland

Primary Maths for Scotland



# 1st Level Maths
## Problem Solving Pack

Authors: Sheona Goodall, Karen Hart, Linda Lapere
Series Consultant: Carol Lyon
Series Editor: Craig Lowther

001/17082002

10 9 8 7 6 5 4



ISBN 9780008508678

Published by
Leckie
An imprint of HarperCollinsPublishers
1 Robroyston Gate, Glasgow G33 1JN

T: 0844 576 8126 F: 0844 576 8131

leckiescotland@harpercollins.co.uk www.leckiescotland.co.uk

HarperCollins Publishers
Macken House, 39/40 Mayor Street Upper, Dublin 1, D01 C9W8, Ireland

Publishers: Fiona McGlade and Jennifer Hall
Project editor: Peter Dennis

Special thanks
Copy editor: Louise Robb
Illustrations: Ann Paganuzzi
Layout: Siliconchips Services Ltd

A CIP Catalogue record for this book is available from the British Library.

Acknowledgements
Images © Shutterstock.com

Whilst every effort has been made to trace the copyright holders, in cases where this has been unsuccessful, or if any have inadvertently been overlooked, the Publishers would gladly receive any information enabling them to rectify any error or omission at the first opportunity.

Printed in Great Britain by Ashford Colour Ltd.

FSC www.fsc.org MIX Paper | Supporting responsible forestry FSC® C011748

# Contents



**Download free resources at www.collins.co.uk/primary-maths-for-scotland-free-resources**

# Introduction

## Primary Maths for Scotland overview

Primary Maths for Scotland comprises a whole-school mathematics programme, which also includes the Scottish Curriculum for Excellence Early and Second Levels. The Primary Maths for Scotland series is designed to develop conceptual understanding and procedural fluency, and can be easily implemented alongside existing plans. It has been written by a team of Scottish Numeracy and Mathematics specialists, especially for Scottish schools, and has been aligned to the Scottish Curriculum for Excellence Benchmarks to plan routes through the Experiences and Outcomes for Numeracy and Mathematics.

While this First Level Problem Solving book complements the First Level Primary Maths for Scotland Teacher Guide, Textbooks and Assessment Packs, it can also be used as a standalone resource. Research-informed pedagogy was used to guide the creation of these resources. The theory of social constructivism underpins the approach and has informed the phrasing of resources and teaching ideas.

Pupils should be active participants in the learning process. They should be encouraged, and challenged, to talk about their ideas and strategies for solving mathematical problems. We want all pupils to enjoy learning mathematics. This approach supports positive learning experiences and the development of metacognition.

## Problem Solving

In his seminal work, Polya (1945) articulated the process of solving problems as:

1. Understand the problem
2. Devise a plan
3. Carry out the plan
4. Examine the solution obtained (look back and reflect)

Since the publication of Polya's work, our understanding of problem solving and its importance to learning, life and work has grown considerably. Problem solving is far more than a process; it incorporates thinking skills, attitudes towards learning and strategies for solving problems. As stated in the *Mathematics Principles and Practice* paper, "Mathematics is important in our everyday life, allowing us to make sense of the world around us and to manage our lives. Using mathematics enables us to model real-life situations and make connections and informed predictions. It equips us with the skills we need to interpret and analyse information, simplify and solve problems, assess risk and make informed decisions." (Scottish Government, 2011)

The ability to solve problems is vital to every child and young person as they grow and develop. In addition to supporting pupils' understanding of the curriculum, good problem-solving behaviours allow them to make sense of their world, preparing them for their future lives and for the world of work. Indeed, in many high-performing education systems, problem solving is considered the goal of the mathematics curriculum. In these systems, problem solving is at the heart of the curriculum, building not only on sound mathematical processes, skills and understanding of concepts, but also on collaboration, communication, metacognition and a positive attitude to learning mathematics.

This Primary Maths for Scotland problem-solving resource has been written with all of the aforementioned aspects in mind. At Leckie, we are strong advocates of a positive and enjoyable mathematical experience. Pupils who struggle with mathematics need to experience success in order to rebuild their confidence. It is therefore vital that teachers ensure that pupils have the mathematical *knowledge* required to access the problem, so that they can focus on *how* they might solve it.

An important part of the problem-solving process is allowing pupils space to *puzzle*, space to try out and discuss ideas, and space to be creative in their response to the problem posed. We do not advocate leading pupils through a problem, nor introducing a specific strategy in advance of the problem-solving activity. To do so can reduce pupil agency, curtail opportunities to link previous learning to this new situation, and stifle creativity.

Through the process of solving problems, pupils will develop a range of transferable strategies, methods and skills. These should be celebrated and reviewed in order to scaffold the pupils' thinking and encourage metacognition.

## Linking with the Numeracy and Mathematical Skills

> "To face the challenges of the 21st century, each young person needs to have confidence in using mathematical skills, and Scotland needs both specialist mathematicians and a highly numerate population." (Scottish Executive, 2006)

The Benchmarks for Numeracy and Mathematics (Education Scotland, 2017) outline eight overarching skills, embedded within the Experiences and Outcomes. By following the advice given in this resource, teachers can support the development of these skills. The following notes how each skill can be developed through engagement with the Primary Maths for Scotland problem-solving tasks.

- **interpret questions**. For every problem the pupil will need to work out what is being asked of them and decide how to proceed. Some will need to ask questions to clarify their thoughts or to simplify the problem in their own mind.
- **select and communicate processes and solutions**. The pupils will need to choose a starting point and be challenged to justify their choice. As they work on the problems, they should be encouraged to share their thinking. This could be through a discussion, by sharing a picture, graph or table or by using concrete materials to model or act out the situation. All of these processes are enhanced when pupils are encouraged to share and talk about their strategies.
- **justify choice of strategy used**. The *Let's discuss* section suggests a range of possible strategies that pupils might use. In addition to sharing strategies and ideas as they work on a problem, teachers should initiate a whole group plenary in which the relative merits of the different strategies used by the class/group are discussed. By asking the pupils to draw comparisons between the methods used, and with methods that they have encountered previously, we can support metacognition.
- **link mathematical concepts**. Many of the problems use mathematical concepts from across the maths curriculum. Being able to link concepts, and connect different areas of maths together, will help pupils experience success and have a positive impact on their confidence.
- **use mathematical vocabulary and notation**. The problems in this pack cover almost all of the Experiences and Outcomes at First Level. As such, there will be many opportunities for the pupils to practise using key vocabulary and notation across the mathematics curriculum.
- **use mental agility**. There are opportunities for the pupils to decide how to calculate, make jottings, and select from and use a range of mental strategies (see Experiences and Outcomes identified against each problem)
- **reason algebraically**. There are opportunities for the pupils to look for patterns, use symbols and make generalisations (see Experiences and Outcomes identified against each problem)
- **determine the reasonableness of a solution.** For every problem we suggest that pupils are challenged to justify their answers, to see if their answers are reasonable in the context of the problem, and to work collaboratively to check their solutions.

## The Problems

Each problem in the resource has the following structure to help you plan and carry out the activity with your class or group.

## Experiences and Outcomes

### Experiences and Outcomes

Number and Number Processes

- *I can use addition, subtraction, multiplication and division when solving problems, making best use of the mental strategies and written skills I have developed.* ***MNU 1-03a***

We have identified the Experiences and Outcomes that are directly linked to the problem. This is usually through the mathematical skills and knowledge required for the problem. It may be possible that the pupils end up making links with other areas of the mathematics curriculum not identified. This is the nature of problem solving.

## Resources

### Resources

These should be available for pupils to use if they wish:

- pencil and paper
- number lines or hundred squares

This section lists the physical resources, interactive whiteboard visuals and worksheets that should be available to support the problem-solving process. Where visuals or worksheets are required, we have provided these as downloadable resources (in colour).

Go to **www.collins.co.uk/primary-maths-for-scotland-free-resources** to download your free resources.

At all times we recommend that resources such as scrap paper, concrete materials, measures, rulers etc. are available for the pupils to use if they wish. We do not issue equipment, rather we wait to see what approach the pupils want to try first. Some will want to talk through the problem, discuss and pose ideas and questions. Some pupils will want to try a few calculations on paper or with a calculator, or perhaps draw a picture or diagram. Some will want to act out the situation, sometimes using concrete materials. Having the materials to hand allows this freedom of choice.

We have listed resources that we use regularly with our own classes. We do not expect you to buy specific items of equipment. Many objects, such as counters or cubes, will be available from different retailers. Items from the outdoors (for example, pebbles) can also be used.

## Before they start

### Before they start...

Pupils should be able to add one and two-digit whole numbers within 50. It may be beneficial to discuss the word 'consecutive'.

We have listed the specific skills, knowledge or processes that pupils will need to access the problem. The pupils should be familiar with these and, if possible, the context of the problem before they start. This frees up cognitive space for the pupil to focus on the problem in hand. The aim is to develop problem-solving skills, attitudes and effective communication.

There will be occasions when pupils are struggling to remember the name of a shape or a process for calculating. Our advice in these situations is to facilitate the problem-solving aspect. Please do provide calculators, aide memoirs, access to posters and other easily accessible information if this helps the pupils to stay on task with the problem.

## Let's go - enabling prompt and extension activity

### Let's go!

You have four jars and 16 sweeties. Put the sweeties in the jars so that each jar has an odd number of sweeties. **How many different combinations can you make?**

### Enabling prompt

- What if you have eight sweeties and four jars? Share the sweeties between the jars so that each jar has an odd number. What could you use to help you?
- What if the first jar has five sweeties and the second jar has five sweeties?

The **Let's go** section contains the problem to be solved. It is important that pupils are allowed to puzzle initially with minimal, and on occasion no, input from the teacher. This will support deeper discussion and support a focus on metacognition as pupils progress through the problem. Some problems are open-ended in nature and have multiple solutions; others have a definite answer.

Creativity is at the heart of problem solving. Even problems with a clear solution provide opportunities for creativity. As well as using the relevant mathematical skills, knowledge and processes, pupils should be encouraged to experiment, explore, investigate and reflect on their journey towards reaching a solution.

Some pupils will struggle to engage and get started. For these pupils, we have created **Enabling prompts**. Enabling prompts may include questions to stimulate thinking and discussion or suggest ways to simplify the original problem. It may be that the pupil does not have the mathematical knowledge required for the problem, perhaps they lack resilience or confidence to try out something new. It is important that the pupil is allowed to engage with the problem at their own 'level' and is not left to drift. Providing a simpler example, peer help or support from a member of staff can make the problem accessible to these pupils and ensure that they feel included.

Some pupils will solve the problem quickly, demonstrating mastery of the task. The **Extension activities** can be used to stretch pupils who solve the problem quickly. 'What if….' type questions can also be used to challenge these pupils and deepen their understanding.

## Let's discuss

### Let's discuss!

It is likely that pupils will begin by randomly plotting routes through the maze and totaling the numbers. Some may then try to adjust their route to add in higher numbers or lower numbers depending on whether their total was more than or less than 100. This, or other strategic approaches, should be shared at appropriate points. Pupils should be encouraged to share their approaches to recording their routes and checking that they were not repeating the same route. Some discussion of the desired total of 100 and the size of the numbers within the cells would be worthwhile. The teacher might ask **Roughly how many cells do you think your route might need to go through to get close to 100? What makes you think that?**

In the **Let's discuss** section we have provided a narrative on some of the possible strategies that pupils may use to solve the problem. These can be used in a number of ways: to provide prompts to help the pupils; to pose questions to the pupils to deepen their understanding; or to help you recognise when pupils are either on the right track or searching down a blind alley. Please note that this is not an exhaustive list of expected strategies, nor a directive that pupils must follow these routes to the solution. In problem solving there are many, many strategies that can be used. What we have found is that pupils will develop a range of problem-solving skills that they will come to rely on. The challenge is to support them to grow these strategies and to see if they can apply them to new and unfamiliar contexts.

## Let's check

### Let's check!

Enabling prompt: The five animals are three chickens (six legs) and two sheep (eight legs).

Main problem: There are 11 chickens (22 legs) and 7 sheep (28 legs).

Extension activity: There are many possible solutions for this. These are two possibilities:

| **Bicycles** | **3-wheeled off-road vehicles** | **4-wheeled tractors** | **Total number of wheels** |
|---|---|---|---|
| 3 (6 wheels) | 4 (12 wheels) | 8 (32 wheels) | 50 |
| 4 (8 wheels) | 6 (18 wheels) | 6 (24 wheels) | 50 |

The **Let's check** section provides suggested solutions to the Let's go problem, the Enabling prompt and the Extension activity, except for situations where the problem is more open ended and there are multiple solutions. In this case we have tried to give an example of the possible solutions to look out for. We advocate that the pupils have an opportunity to share their solutions with their peers before the teacher confirms if an answer is correct. When a pupil claims to have solved the problem, you can prompt for understanding by posing questions such as: **How did you solve the problem? Have you checked your solution? Have you compared your solution with another pupil? Can you explain your method to a partner? Does your method work for all the situations in the problem? can you make your thinking / solution clearer for all to see?**

## Let's reflect

### Let's reflect!

Pupils should think about how they needed to persevere when trying various things as they worked towards solutions in this problem. The teacher may ask **How did it feel when you found something that worked for one pair of numbers but then discovered that it did not work for all of the pairs?** Pupils should also be encouraged to reflect on how incorrect ideas helped them to work towards solutions.

To promote metacognition, the **Let's reflect** section provides specific guidance for each problem task. As pupils become more experienced with solving problems, they can be prompted to think back to similar situations, to see if they can remember and apply a strategy that they have used successfully in the past.

## Teacher Guide Reference

**Teacher Guide Reference**

- First Level Teacher Guide Units 7 and 12

While solving the problem, you may notice gaps in the pupil's mathematical knowledge or skills. These links will help you to plan revision for these missing skills.

# Planning and Carrying out the Problem Solving Activity – Lesson Planning Advice

The following is a series of lesson planning prompts, based on our experience of facilitating problem solving with our own pupils.

## Planning and Preparation

Prior to commencing, check that the pupils have the mathematical knowledge necessary to access the problem, and that they understand the vocabulary involved.

Active differentiation is supported by the use of the Enabling prompts and Extension activities. Pupils' work will also be differentiated by outcome as they will use a variety of strategies and record their solutions with differing levels of complexity.

Make any physical resources readily available.

Plan key questions. The following, or similar, question stems can be prepared in advance ready to display:

What are you trying to find out?

How are you going to start?

How are you going to solve the problem?

What information do you have?

What information do you need?

What have you found out so far?

Can you show me a different way?

Can you tell me more about ..?

What would happen if ..?

What can you change?

Is there anything you can't change? Why is this?

Where can you find out more information?

Why do you think that?

Is that always true? Tell me why you think that.

How do you know ..?

What else could you do?

How did you check your solution?

Tell me about a similar situation to this.

Explain your strategy

Explain how you found your solution

## Presenting the Problem, Managing the Learning

*The activity in a nutshell*

The problem is introduced to the whole class or group. Pupils then move into smaller groups, trios or pairs or, if appropriate for the child, work on their own. After 5-10 minutes the teacher brings the class / group back together to discuss their initial ideas before returning the pupils to their smaller groups to complete the problem. Finally, the whole class/group comes together again to reflect on and discuss the approaches used.

*The physical space*

You will need to plan how the pupils will use the space available. Can all pupils see the visuals and questions displayed or do you need extras to share around? Will the pupils start as a large group at the front of the class and then move back to their tables, or is there another arrangement that you and the pupils like? Does the space allow for side-by-side working that can quickly become a face-to-face discussion space? It is important to plan for movement and transitions as attention to the task can be lost if there is too much disruption.

*Getting started – the first 10 minutes*

Your ultimate aim, as you start the **Let's go** section, is to grab each pupil's attention and stimulate their interest so that they become enthused and motivated to engage with the task. Some problems have a real-life context, some are fantasy contexts and others are in a mathematical context. The problem may present a situation that pupils are familiar with, or an unfamiliar context to pique their interest. It could be a counter intuitive situation that they want to disprove, or a situation that is *current* for them. If you feel that changing the context would help engage your pupils, then do so. Grabbing the pupils' attention and achieving engagement with any task is arguably the most important aspect of the classroom experience. Allowing for personalisation and choice is integral to the problem solving process. Many pupils will see the activity as 'play' and we encourage this view of mathematics as an enjoyable and creative experience. Remember to let the pupils have a go before any intervention.

*Moving on – the middle of the lesson*

After this initial phase, bring the group together to share their ideas on how to solve the problem and allow them to ask questions. It is helpful to have prompt questions ready to stimulate discussion. The pupils should by now have devised a plan to solve the problem. They can then return to their small groups to try out their ideas.

Most pupils will be able to follow their plan based on their prior experiences and through discussion with a partner or with the group. Use the **Enabling prompts** with pupils who are struggling.

Throughout the activity remember that relevant equipment and resources should be available for the pupils to use. Some children may be hesitant to use concrete materials or to act out the situation. Encourage them to use these resources if it helps them visualise the problem and make progress.

*Solving the problem*

As the pupils start to create solutions they will want to know if their answer(s) are correct. Some problems have definitive answers, others are more open ended. Develop a range of key, 'coaching questions' (for example who, what, where, when, why and how question stems) and have these on display. Modelling the use of such questions can encourage pupils to use similar questions in their own discussions. Challenge the pupils to justify why they think their solution is correct. Encourage them to check their answers by estimating or recalculating, compare their solution(s) with a partner and explain their strategies to their peers. Finally, you can reveal the solutions once you are satisfied that the pupils have developed self-checking and estimation skills.

*Extending the problem*

Some pupils will quickly solve the problem, demonstrating a mastery of the task. At this point you can use further questioning to deepen their understanding. How have they communicated their solution? Are they organised? Did they use a picture, diagram or table? The **Extension activities** can be used to stretch the pupils to apply their strategies again at a higher level of challenge.

*Ending the problem*

This time is to promote metacognition. Bring the group or class together and have questions ready that you can use to prompt reflection on the task. For example, **Have you done a task like this before? What strategy did you use to solve the problem? Have you used this strategy before?** 'Think, Pair, Share' is a useful strategy, allowing pupils time to organise their thoughts, rehearse their response with a partner and then say it out loud to the whole group. There are reflective questions given for each task in the **Let's reflect** section.

# References

Education Scotland (2017) *Benchmarks Numeracy and Mathematics available* at https://education.gov.scot/nih/Documents/NumeracyandMathematicsBenchmarks.pdf

Polya, G (1945) *How to Solve it*. Princeton University Press

Scottish Executive (2006) *Building the Curriculum 1: The Contribution of Curricular Areas*. Edinburgh: Scottish Executive. https://education.gov.scot/Documents/btc1.pdf

Scottish Government (2011) *Mathematics Principles and Practice*. Edinburgh: Scottish Government. https://education.gov.scot/Documents/mathematics-pp.pdf

# Bob's Birthday

## Experiences and Outcomes

Number and Number Processes

- *I can share ideas with others to develop ways of estimating the answer to a calculation or problem, work out the actual answer, then check my solution by comparing it with the estimate.* ***MNU 1-01a***
- *I can use addition, subtraction, multiplication and division when solving problems, making best use of the mental strategies and written skills I have developed.* ***MNU 1-03a***

## Resources

These should be available for pupils to use if they wish:

- blocks/counters/lolly sticks/straws to represent candles
- Resource sheet: Bob's Birthday Blank Cakes
- Resource sheet: Bob's Birthday Estimation Cake

## Before they start...

Pupils should understand the term 'estimate' and should be able to add single and double-digit numbers.

## Let's go!

Ensure pupils can visualise the scenario and that they understand what the problem is asking before they estimate and record an answer.

Bob, the school cat, gets a birthday cake made from cat food every year! The children blow out the candles for Bob. Nuria brings a cake she has made.

'Happy Birthday, Bob', she says. 'There are 7 candles on the cake because you are 7 years old! When you were 6 you had 6 candles.'

'I wonder how many candles we have used since Bob was born?' said Amman.

Can you estimate the answer? Write it down. Now work out the exact answer and show your thinking.

# Enabling prompt

If pupils are unable to get started on the problem, or do not understand the problem, one of the following prompts may help:

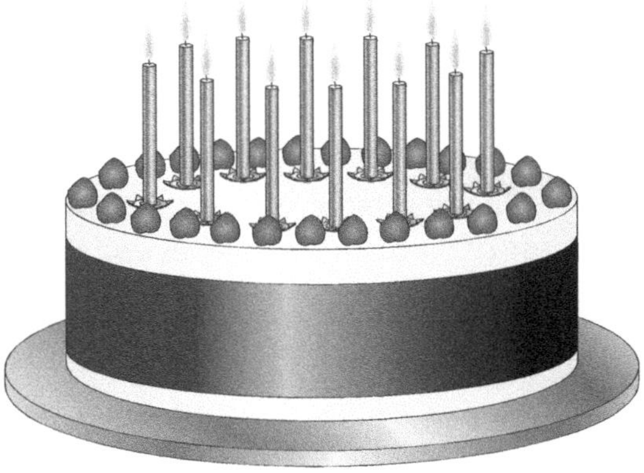

- Without counting, how many candles would you estimate there are on this cake? More or less than 5? More or less than 10?
- Work out how many birthdays Bob has had.
- Use concrete materials to represent candles.
- How many candles were used when Bob was 1?
- How many had been used when Bob was 2?
- Write a number sentence.

# Extension activity

- How old will Bob be when the children have used 45 candles?
- Nuria thinks they will have used 70 candles on Bob's 11th birthday. Is she correct?
- Can you see a pattern in your answers? Can you keep the pattern going?
- Think of a different way to show your thinking.

## Let's discuss!

When pupils have estimated an answer, the teacher should pause and discuss estimations. It may be that the teacher asks **Do you estimate there will be more than 100 candles?** or **Do you estimate that there will be less than 20 candles?** At this stage, all estimates should be listed on the board and pupils should have the opportunity to explain their reasoning. Pupils should keep the estimation and resist changing it to an exact answer later.

Teachers can facilitate discussion by using a prompt such as:

- **How can estimates help us when we think we have an answer?**

Teachers may identify children who display little number sense of larger numbers.

When pupils have had an opportunity to solve the problem, teachers should gather methods that they have used to record their thinking. The focus should be on the second part of the problem, 'Show your thinking'.

Teachers can facilitate discussion by using prompts such as:

- **Did you notice any patterns?**
- **Did you find any pairs to help you add?**

Some pupils may use concrete or pictorial representations as they work on this problem. They should be encouraged to share and explain their representations with the class, and attention should be given by the teacher to how pupils recorded the process rather than on the solution alone. Teachers should draw attention to those who represent the problem without numbers and those who represent the problem and include numbers. They may ask **Are the numbers useful?**

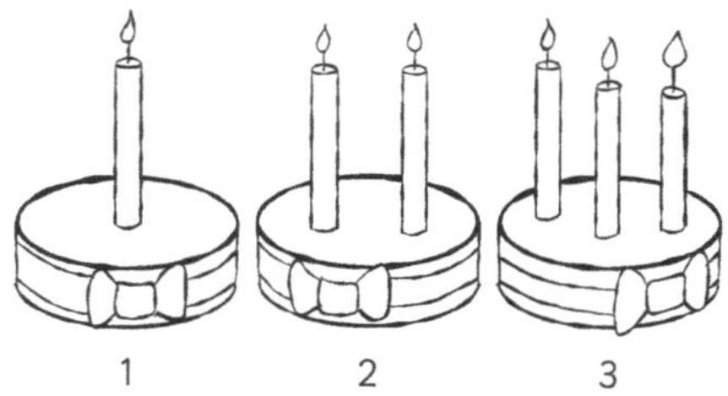


Some pupils may recognise the problem as 1 + 2 + 3 + 4 + 5 + 6 + 7 = 28

Attention can be drawn to how looking for pairs of numbers that add to ten can make adding easier. This is also an opportunity to introduce the vocabulary 'consecutive numbers' if teachers feel it is appropriate.

After the discussion, pupils may wish to explore further using the suggestions in the Extension activity.

## Let's check!

**Main problem:**

There would have been 28 candles in total.

**Extension activity:**

Bob would be 9 when 45 candles have been used.

Nuria is wrong. The children will have used 66 candles by Bob's 11th birthday.

## Let's reflect!

When pupils have had an opportunity to discuss their methods and solutions, teachers may wish to facilitate discussion around the information that pupils focused on in the initial stages of the problem. Teachers may ask **What did you do to help them 'see' the problem? How do you feel when an estimate and an answer are different? What is the purpose of an estimate?**

Teachers can draw attention to the use of estimation as a rough guide as to whether the answer is likely or not.

## Teacher Guide Reference

- Early Level Teacher Guide Unit 1
- First Level Teacher Guide Unit 3

# The Abacus

## Experiences and Outcomes

Number and Number Processes

- *I have investigated how whole numbers are constructed, can understand the importance of zero within the system and can use my knowledge to explain the link between a digit, its place and its value.* ***MNU 1-02a***

## Resources

These should be available for pupils to use if they wish:

- an abacus (may be real or a virtual tool), or a printed, laminated copy of an abacus
- paper and pencil for jottings
- Resource sheet: The abacus

## Before they start...

Pupils should understand the place value associated with each digit in three-digit whole numbers.

## Let's go!

The children are playing with some abacuses in the Maths area in their classroom. They each have one abacus and 10 beads.

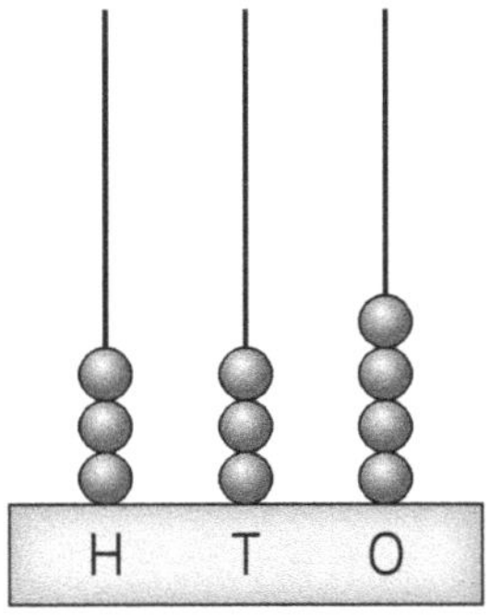

Isla says, 'Look, I have made two hundred and fifty three on my abacus'.

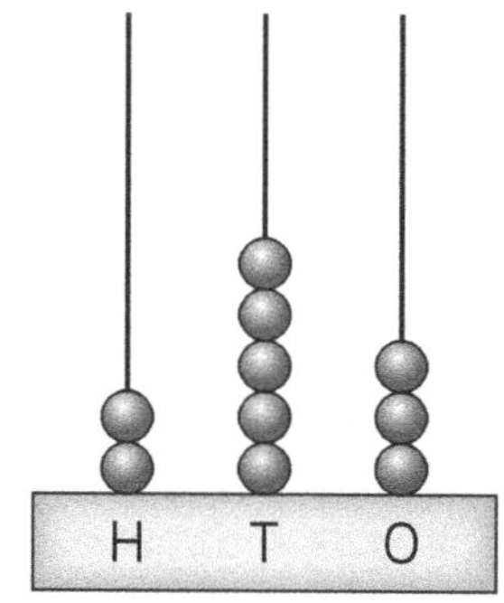


Amman wonders what the highest number is that can be made on this abacus if all 10 beads are used and each rod has at least 1 bead on it.

**What do you think the highest number will be?**

Finlay decides to make the lowest number that can be made on this abacus using all 10 beads, making sure that each rod has at least one bead on it.

**What number does he make?**

Nuria wants to use all 10 beads to make a number that has roughly the same number of beads on each rod. What numbers might she make?

## Enabling prompt

Pupils who are struggling to make a start to this problem may be given this simpler version. The teacher may alter the number of beads to suit the needs of the pupil.

This abacus shows the number 25 using seven beads. What different numbers can you make on this abacus using seven beads? Each rod must have at least one bead on it.

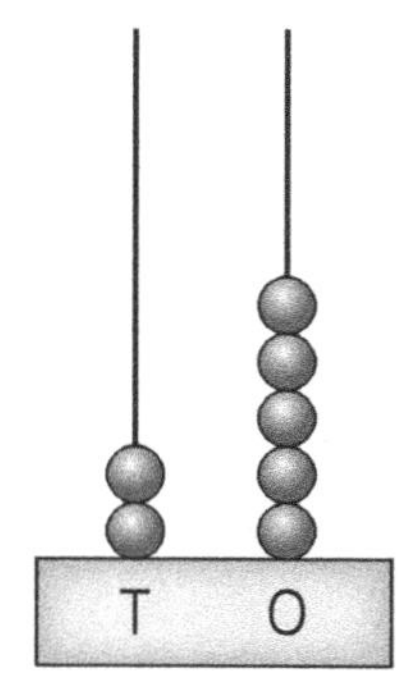


## Extension activity

The children find a different abacus set. There are 20 beads altogether in this set. Write down all of the numbers that can possibly be made with this abacus set if each rod has at least one bead on it.

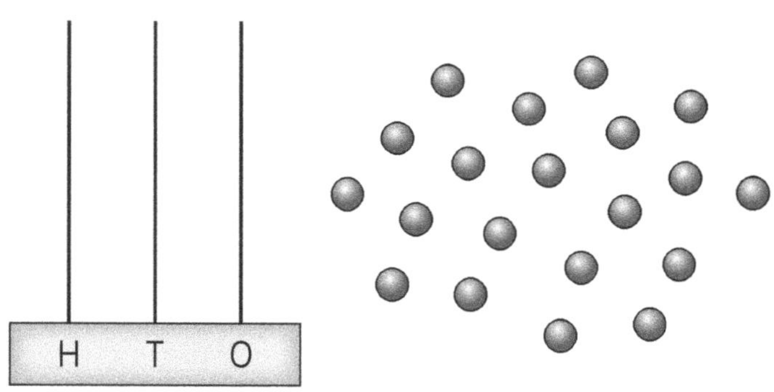

## Let's discuss!

This problem provides good opportunities for pupils to work in pairs and trios, discussing the problem with each other. Initially, pupils should be left to discuss Isla's number (253) and work out how the abacus represents this. The teacher should then facilitate a class discussion of how numbers can be represented on an abacus to ensure that all pupils are clear about the value associated with each digit. Pupils should then work together in their pairs or trios to find the highest number, the lowest number and a number that has roughly the same number of beads on each rod. Some pupils will work with only digits while others will use diagrams to represent an abacus, or an actual / virtual abacus if this is available. In directing pupils to check solutions and sharing pupils' thinking, the teacher should use actual or virtual manipulatives wherever possible within demonstrations. The teacher may use prompts such as **Where should I put this bead if I want it to have the highest value?** (holding a bead, or pointing to a bead drawn on paper) and **I wonder which rod will have the most beads on it for the highest number?** and so on. Where a group presents a number that is not the highest or lowest, the teacher may say something like **What number could I get if I move this bead onto a different rod?** (pointing to an appropriate bead).

The Extension activity introduces opportunities for discussion of the highest number of beads that can go onto a single rod. The teacher might create a number that has 9 ones or 9 tens and then say **What will I get if I put another bead onto this rod?** (pointing to the ones or the tens as appropriate). This should encourage discussion of 1 ten being equivalent to 10 ones, 1 hundred being equivalent to 10 tens, and so on.

## Let's check!

**Enabling prompt:** 16, 25, 34, 43, 52, 61 can be made.

**Main problem:**

Amman makes 811 as the highest number.

Finlay makes 118 as the lowest number.

Nuria might make 433, or 343, or 334.

**Extension activity:** there are 36 different numbers, as follows:

992, 983, 974, 965, 956, 947, 938, 929,

893, 884, 875, 866, 857, 848, 839,

794, 785, 776, 767, 758, 749,

695, 686, 677, 668, 659,

596, 587, 578, 569,

497, 488, 479,

398, 389

299

## Let's reflect!

Pupils should be encouraged to think about how discussing this problem with their peers helped them all to make progress. The teacher may ask **In what way did your discussion help you to make progress? How did you encourage each other?** They should also think about how being methodical was beneficial. The teacher may ask **How can you be sure that you have the highest / lowest number? How did you make sure you did not miss any numbers? Did you find it easier when you were organised in the way you wrote things down?**

## Teacher Guide Reference

- First Level Teacher Guide Unit 2

# Animals on the Farm

## Experiences and Outcomes

Number and Number Processes

- *I can use addition, subtraction, multiplication and division when solving problems, making best use of the mental strategies and written skills I have developed.* ***MNU 1-03a***

## Resources

These should be available for pupils to use if they wish:

- Resource sheet: graduated or empty number lines

## Before they start...

Pupils should be able to add and subtract whole numbers within 50. Being able to work with multiples of 2 and 4 will also be beneficial.

### Let's go!

Isla and Amman visit a local farm. Out in the field are chickens and sheep. Isla and Amman can see 18 animals and 50 legs. How many chickens and sheep are in the field?

## Enabling prompt

If pupils are unable to get started on the problem, or do not understand what the problem requires them to do, one of the following prompts may help:

- How many legs would there be if there was one chicken and one sheep?
- In another field, Isla and Amman can see 5 animals and 14 legs. How many chickens and sheep can they see there?

## Extension activity

Nuria and Finlay are in one of the barns, which is full of bicycles, three-wheeled off-road vehicles and four-wheeled tractors. They see 50 wheels. **How many of each vehicle could there be?**

## Let's discuss!

Giving pupils time to independently puzzle over the task, try different strategies in order to begin the task, make errors and self-correct is important. Teachers may wish to gather learners together *briefly* after approximately 10 minutes to share strategies/methods being used. Teachers may ask **What have you noticed? What have you tried? How are you recording your answers?** Allow time for pupils to share how they approached the task, what they have tried and if there were patterns that were spotted.

Some pupils may use 'Guess and Check', randomly selecting numbers to try to calculate the 50 legs from 18 animals, whereas some may work systematically and adjust their inaccurate answers by reducing or increasing numbers slowly.

Others may choose to draw a table to help record their attempts and spot any patterns using the fact that two chickens have the same number of legs as one sheep. For example, they may have calculated 50 legs but have six sheep and 13 chickens.

Problem solving can be haphazard and pupils should be encouraged to record their thinking. If pupils are becoming frustrated that they cannot recall how they have solved the problem, the teacher can ask them how they can keep a record of what they did. Teachers can gather a great deal of information at this time by observing methods and conversations.

When pupils have had an opportunity to solve the problem, teachers should gather methods that pupils have used to record their thinking. Teachers may then facilitate discussion by using prompts such as:

- **What advice would you give to another class doing this problem?**
- **How did you start this problem?**
- **What helped you solve the problem (if you managed to)? What didn't help?**

## Let's check!

**Enabling prompt:** The five animals are three chickens (six legs) and two sheep (eight legs).

**Main problem:** There are 11 chickens (22 legs) and 7 sheep (28 legs).

**Extension activity:** There are many possible solutions for this. These are two possibilities:

| Bicycles | 3-wheeled off-road vehicles | 4-wheeled tractors | Total number of wheels |
|---|---|---|---|
| 3 (6 wheels) | 4 (12 wheels) | 8 (32 wheels) | 50 |
| 4 (8 wheels) | 6 (18 wheels) | 6 (24 wheels) | 50 |

## Let's reflect!

After pupils have had the opportunity to discuss their methods and solutions, teachers may wish to facilitate discussion using questions such as **How did you feel when you first read the problem? Did your feelings change as the time progressed? What helped you persevere when things got tricky?**

## Teacher Guide Reference

- First Level Teacher Guide Unit 3

# Consecutive Numbers

## Experiences and Outcomes

Number and Number Processes

- *I can use addition, subtraction, multiplication and division when solving problems, making best use of the mental strategies and written skills I have developed.* ***MNU 1-03a***

## Resources

These should be available for pupils to use if they wish:

- pencil and paper
- number lines or hundred squares

## Before they start...

Pupils should be able to add one and two-digit whole numbers within 50. It may be beneficial to discuss the word 'consecutive'.

## Let's go!

The number 3 can be written as a sum of consecutive numbers: 1 + 2 = 3. The number 15 is also a sum of consecutive numbers and this can be written in three different ways: 7 + 8, 4 + 5 + 6 and 1 + 2 + 3 + 4 + 5. Can all numbers between 1 and 20 be written as sums of consecutive numbers? Can some consecutive number sums be written in more than one way?

## Enabling prompt

- What numbers come before 4?
- What do the consecutive numbers 1, 2 and 3 add up to?
- Can you make 5 by using consecutive numbers?
- What is the total of these consecutive numbers: 1 + 2 + 3, 3 + 4? What different numbers can these consecutive numbers make?

## Extension activity

- What do you notice about the total when you add two consecutive numbers?
- What do you notice when you add three consecutive numbers?
- What about five consecutive numbers?
- Are there any other patterns you notice?
- What is the largest number you can create with consecutive numbers?

## Let's discuss!

Giving pupils time to independently puzzle over the task, try different strategies in order to begin the task, make errors and self-correct, is important. Teachers may wish to gather pupils together *briefly* after approximately 10 minutes to share strategies/methods that are being used. Teachers may ask **What have you noticed? What have you tried? How are you recording your answers?** Allow time for pupils to share how they approached the task, what they have tried and if there were patterns that were spotted.

This is a problem that may be tackled either independently or collaboratively.

Some pupils may start by trying to randomly create any number larger than four with consecutive numbers. Some may work systematically by trying to create four, then five, then six and so on.

Others may start by adding consecutive numbers to see which totals they calculate. Again, this may be done randomly by some pupils and systematically by others. For example, starting with 1 + 2 then 1 + 2 + 3 then perhaps moving on to 2 + 3 then 2 + 3 + 4.

Problem solving can be haphazard and pupils should be allowed to find out for themselves that there is a need to record their thinking. If pupils are becoming frustrated that they cannot recall which consecutive numbers they have used, or which numbers they have created, the teacher can ask them how they can keep a record of what they did. Teachers can gather a great deal of information at this time by observing methods and conversations. Attention can also be drawn to the advantages of having a way of recording answers that allows the pupil to know they have explored all possible options.

When pupils have had an opportunity to solve the problem, teachers should gather methods that pupils have used to record their thinking. Teachers may then facilitate discussion by using prompts such as:

- **How did you start this problem?**
- **Were some totals easier to calculate than others?**
- **Did you spot any patterns?**

## Let's check!

**Main problem:**

These are the consecutive numbers that total numbers up to 20.

| Numbers that can be made by adding consecutive numbers | Numbers that cannot be made by adding consecutive numbers |
|---|---|
| 3 = 1 + 2 | 2 |
| 5 = 2 + 3 | 4 |
| 6 = 1 + 2 + 3 | 8 |
| 7 = 3 + 4 | 16 |
| 9 = 4 + 5, 2 + 3 + 4 | |
| 10 = 1 + 2 + 3 + 4 | |
| 11 = 5 + 6 | |
| 12 = 3 + 4 + 5 | |
| 13 = 6 + 7 | |
| 14 = 2 + 3 + 4 + 5 | |
| 15 = 7 + 8, 4 + 5 + 6, 1 + 2 + 3 + 4 + 5 | |
| 17 = 8 + 9 | |
| 18 = 5 + 6 + 7, 3 + 4 + 5 + 6 | |
| 19 = 9 + 10 | |
| 20 = 2 + 3 + 4 + 5 + 6 | |

**Extension activity:**

- All odd numbers can be written as the sum of two consecutive numbers.
- Three consecutive numbers always total a multiple of three.
- Multiples of five are the sum of five consecutive numbers (from 15 onwards)
- Powers of two (2, 4, 8, 16 . . .) cannot be made by adding consecutive numbers.

## Let's reflect!

After pupils have had the opportunity to discuss their methods and solutions, teachers may wish to facilitate discussion using questions such as **Where did you start? Did the first strategy you tried work? Did you think there would be one answer or more than one? How did this impact the way you worked and your feelings about the problem?** If pupils collaborated on this problem, the question **What were the advantages or disadvantages of working with others on this problem?** may be useful.

## Teacher Guide Reference

- First Level Teacher Guide Unit 3

# Gran's Purses

## Experiences and Outcomes

Number and Number Processes

- *I can use addition, subtraction, multiplication and division when solving problems, making best use of the mental strategies and written skills I have developed.* ***MNU 1-03a***

## Resources

These should be available for pupils to use if they wish:

- coins and bags or tubs

## Before they start...

Pupils should be able to add and subtract within 20.

## Let's go!

Nuria is visiting her gran.

'Look at my amazing purses', said Gran. She showed Nuria four purses.

'I have fifteen £1 coins', said Gran.

'I split the coins between these four purses.'

'Now I can give you any amount from £1 to £15', said Gran, 'by giving you one, two, three or four of the purses.'

'Can I have £1?' said Nuria.

Gran gave Nuria one purse.

'Can I have £15?' asked Nuria.

Gran gave Nuria all four purses.

**How many coins were in each purse?**

## Enabling prompt

If pupils are unable to get started on the problem, or do not understand the problem, one of the following prompts may help:

- Try a simpler problem. How might 5 × £1 coins be split between three purses? What amounts can you make? What amounts can you not make?
- Act out the problem using the bags and coins. Count out 15 × £1 coins. What do you already know? How much money is in the first purse? What will the coins in the remaining three purses be worth?

## Extension activity

Gran gave Nuria another purse containing £1 coins and said, 'Now I can give you any amount from £1 to £32 by giving you one, two, three, four or five purses.'

Nuria looked at Gran and said, 'Impossible!'

Who is correct, Gran or Nuria? Explain your answer.

### Let's discuss!

Teachers may need to ensure that pupils understand the wording of the problem before allowing them time to explore their thinking. If pupils are finding it challenging to understand what is being asked, use the first enabling prompt.

When pupils have had an opportunity to solve the problem, teachers should gather methods that pupils have used to record their thinking and facilitate a discussion by using prompts such as:

- **How did you decide the amount to put in your first purse?**
- **How do you know you can make every amount from £1 to £15?**

Some pupils will realise that one purse must contain only £1. Teachers can observe what these pupils do next. There may be an element of trial and error and pupils should be allowed to come to the realisation that working systematically can help, thinking **How can I make £1? £2? £3? etc.**

After the discussion, pupils may wish to explore further using the suggestions in the Extension activity.

## Let's check!

**Enabling prompt:**

5 x £1 coins can be shared between 3 purses as follows:

| Purse 1 | Purse 2 | Purse 3 |
|---|---|---|
| £3 | £1 | £1 |
| £2 | £2 | £1 |

You can give Nuria purses so she has £1, £2, £3, £4 or £5.

**Main problem:** £1, £2, £4, £8

**Extension activity:** Nuria is correct. There is £17 in the fifth purse, meaning Gran cannot give Nuria £16 without removing a coin from one of the purses. Every amount between £17 and £32 can be made.

## Let's reflect!

When pupils have had an opportunity to discuss their methods and solutions, teachers may wish to facilitate discussion using questions such as **What helped you to understand the problem? What maths skills did you find useful? How did you know your answer made sense?** Exploring the Extension activity can support pupils in justifying their thinking and challenging each other.

## Teacher Guide Reference

- First Level Teacher Guide Unit 3

# Magic Square

## Experiences and Outcomes

Number and Number Processes

- *I can use addition, subtraction, multiplication and division when solving problems, making best use of the mental strategies and written skills I have developed.* ***MNU 1-03a***

## Resources

These should be available for pupils to use if they wish:

- scrap paper, whiteboards or jotters to draw a 3 × 3 grid
- Resource sheet: graduated or empty number lines to use if they wish
- Resource sheet: Magic Square 1–9 numeral cards

## Before they start...

Pupils should be able to add and subtract two-digit and single-digit whole numbers.

Teachers may need to have a discussion around the vocabulary within the problem to ensure pupils are clear about the meaning of the words ***digits, sum, column, row*** and ***diagonal***.

The following diagram could be used to help explain the vocabulary 'rows', 'columns' and 'diagonals'.

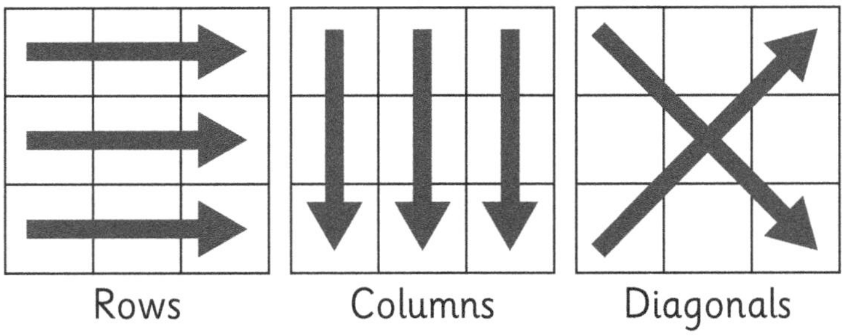

## Let's go!

Place the digits 1, 2, 3, 4, 5, 6, 7, 8 and 9 in the grid so that the sum of the digits in each column, row and diagonal is the same.

| | | |
|---|---|---|
| | | |
| | | |
| | | |

## Enabling prompt

If pupils are unable to get started on the problem or do not understand the problem, one of the following prompts may help:

- What happens when you put the number '5' in the centre?
- What happens if we make this row total 15?
- I have put 9, 6 and 2 in this row. Is that a good choice? How can I make this better?

## Extension activity

- Is this the only way to arrange these 9 numbers?
- Can you create a Magic Square with the numbers 5 to 13? What number would you choose for the centre number? What would the 'magic number' (the total of each row, column and diagonal) be for those numbers? Is it still 15?
- What would happen if the Magic Square was 4 × 4 and you had the numbers 1 to 16?

## Let's discuss!

Giving pupils time to independently puzzle over the task, to try different digit combinations, make errors and self-correct is important. Teachers may wish to gather pupils together *briefly* after approximately 10 minutes to share strategies/methods that are being used, e.g. working backwards. It is important at this point that answers are **not** shared as this is a purposeful discussion around how to get started, sharing what has worked and, importantly, what has not. Encouragement to use number lines or empty number lines may be helpful for pupils who are struggling to accurately recall the number facts they require.

The following questions may help support / challenge pupils:

- **How did you decide where to start?**
- **Can you tell me a bit more about the number in the middle? Why have you chosen that number?**
- **Is there a way of being systematic with the numbers you use? For example, by beginning with the smallest or largest numbers and thinking carefully about their placement.**
- **How do you know that this row totals 13** (for example)? **How can we check?**

Some pupils may respond to the task by placing numbers on the grid to see if they work. Some pupils may benefit from using numeral cards to allow them the freedom to move digits freely. Numbers are then erased (or removed if they have numeral cards) when they realise totals do not match.

Other pupils understand that, within the group of numbers, 9 + 1 = 10, 8 + 2 = 10, 7 + 3 = 10 and 6 + 4 = 10. They use this knowledge to place 5 in the centre. They also realise that they must 'balance out' the large numbers so they are not all on one side of the grid. Pupils may begin by placing 9 + 1, on the grid, then 8 + 2, then 7 + 3.

Some pupils may realise that 1 + 2 + 3 + 4 + 5 + 6 + 7 + 8 + 9 = 45. As there are three numbers in each row, column and diagonal, they divide 45 by 3, which makes 15. They know that the total for each row, column and diagonal must be 15. Pupils may then use one of the other strategies described above to continue with the problem.

When the task approaches a natural end, allow time for pupils to share how they approached it, what they tried and what patterns they spotted. As there are eight answers, it can be helpful to collect different responses from pupils and ask **Are these the same answers or different answers? What makes you think that?**

Ask **What do you notice about the way the numbers are organised in each of the answers?** to support pupils to think about the **symmetry** and **rotation** in the way the numbers are organised in each of the grids.

Ask **What was the 'Magic Number' (the total of every row, column and diagonal) in this task?** (15). **How did you come to realise this was the Magic Number?**

## Let's check!

**Main problem:**

| | | |
|---|---|---|
| 8 | 1 | 6 |
| 3 | 5 | 7 |
| 4 | 9 | 2 |

| | | |
|---|---|---|
| 6 | 1 | 8 |
| 7 | 5 | 3 |
| 2 | 9 | 4 |

| | | |
|---|---|---|
| 4 | 3 | 8 |
| 9 | 5 | 1 |
| 2 | 7 | 6 |

| | | |
|---|---|---|
| 2 | 7 | 6 |
| 9 | 5 | 1 |
| 4 | 3 | 8 |

| | | |
|---|---|---|
| 2 | 9 | 4 |
| 7 | 5 | 3 |
| 6 | 1 | 8 |

| | | |
|---|---|---|
| 4 | 9 | 2 |
| 3 | 5 | 7 |
| 8 | 1 | 8 |

| | | |
|---|---|---|
| 6 | 7 | 2 |
| 1 | 5 | 9 |
| 8 | 3 | 4 |

| | | |
|---|---|---|
| 8 | 3 | 4 |
| 1 | 5 | 9 |
| 6 | 7 | 2 |

## Let's reflect!

When pupils have had an opportunity to discuss their methods and solutions, teachers may wish to facilitate discussion around the merits of certain methods over others. Children should be encouraged to discuss their resilience and confidence when faced with this problem. Teachers may ask questions such as:

- **Did you find the correct answer on your first try?**
- **Did your first idea/method work? Why not/Why?**
- **Did you have the same feelings throughout this task or did your feelings change? When did this happen and why do you think this was?**
- **Do you prefer to tackle problems alone or with someone else?**
- **Did the class discussion help or hinder? Did it change your thinking in any way?**
- **If you were stuck, what helped you get 'unstuck'?**

These questions can offer reassurance to pupils and give teachers useful information to support pupils.

## Teacher Guide Reference

- Early Level Teacher Guide Unit 3
- First Level Teacher Guide Unit 3

# Numbers in a Triangle

## Experiences and Outcomes

Number and Number Processes

- *I can use addition, subtraction, multiplication and division when solving problems, making best use of the mental strategies and written skills I have developed.* ***MNU 1-03a***

## Resources

These should be available for pupils to use if they wish:

- pencil and paper
- number lines or hundred squares
- Resource sheet: Numbers in a Triangle Digit Cards if pupils wish to physically move digits around the triangle
- Resource sheet: Numbers in a Triangle
- Resource sheet: Numbers in a Triangle – Extension activity

## Before they start...

Pupils should be able to add one-digit whole numbers within 20.

## Let's go!

Place the numbers 1 to 6 in each of the circles. The sum of each side of the triangle should equal the number in the centre.

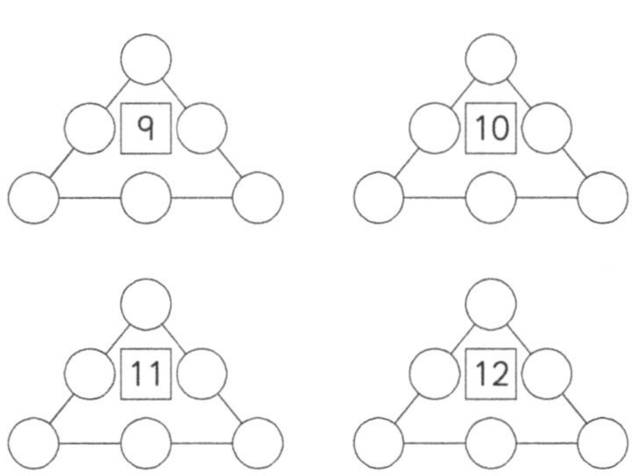

## Enabling prompt

- Can you find three numbers that add up to 9? How could this fit into the '9' triangle?
- What if the corner numbers in the '9' triangle were 1, 2 and 3? Can you complete each side?

## Extension activity

- Using the numbers 1 to 9, create a triangle with 4 numbers on each side so that the totals on the three sides are consecutive numbers.
- Now find another consecutive total triangle.

## Let's discuss!

Giving pupils time to independently puzzle over the task, try different strategies in order to begin the task, make errors and self-correct is important. Teachers may wish to gather learners together *briefly* after approximately 10 minutes to share strategies / methods that are being used. Teachers may ask **What have you noticed about the numbers in the centre and the numbers you have to use? What have you tried? How are you recording your answers?** Allow time for pupils to share how they approached the task, what they have tried and if there were patterns that were spotted.

This is a problem that may be tackled either independently or collaboratively.

Some pupils may use their knowledge of odd and even numbers to reduce the number of combinations in the triangles. Teachers may ask questions such as **How many odd and even numbers do you have to work with? How do you make 9 (an odd number) using odd and even numbers? *(odd + odd + odd or even + even + odd)* Do you make 10 using the same combination of odd and even numbers?**

Some pupils may choose to randomly place digits or digit cards to try to create the target numbers, and use addition and subtraction strategies. It may be worth asking **How are you deciding which numbers to use? How do you place your numbers? How do you know you are correct?** Questions such as these may tease out the 'balancing' of larger and smaller numbers in the triangle.

Others may notice that the numbers in the corners of the triangle are key to creating the totals and the numbers that are in the corners increase as the target number increases. Teachers may ask **What are you noticing about the numbers in the corners compared to your target number? Where are you placing the larger numbers when the target number is 9?**

## Let's check!

**Main problem:**

Possible solutions to the main problem are shown below or may be a rotation of these:

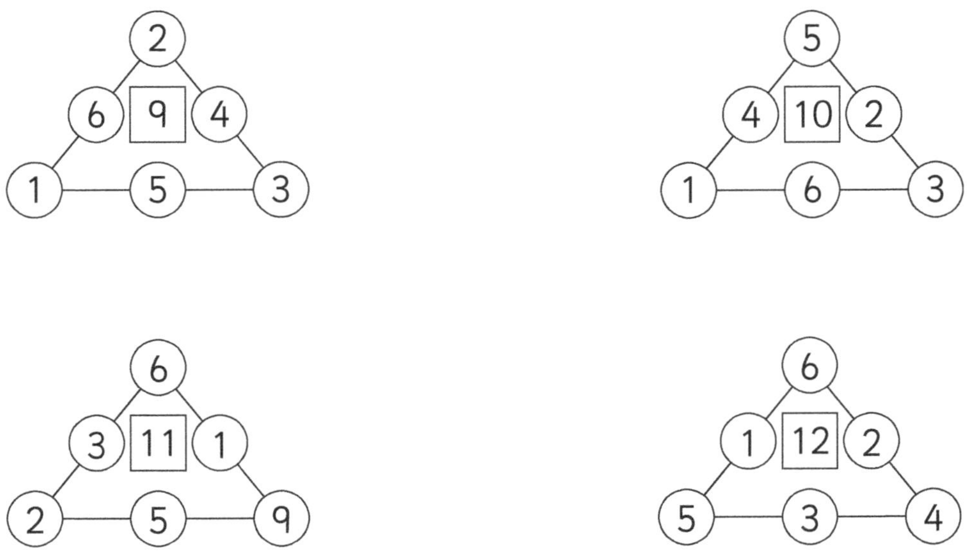


**Extension activity:**

## Let's reflect!

After pupils have had the opportunity to discuss their methods and solutions, teachers may wish to facilitate discussion using questions such as **How did you feel when you first read the problem? Did your feelings change as the time progressed?** The open-ended nature of this problem may be unsettling for some pupils. It may be useful to discuss their feelings around these types of problems. Questions such as **Do maths questions always have just one answer?** may be helpful. If pupils collaborated on this problem, the question **What were the advantages or disadvantages of working with others on this problem?** can be asked. It may also be worthwhile reflecting on their methods – was there an advantage to randomly placing numbers compared to looking at odd/even numbers or balancing out large and small numbers? **Were some number triangles easier to calculate than others? Why do you think this was?**

## Teacher Guide Reference

- First Level Teacher Guide Unit 3

# Pets

## Experiences and Outcomes

Number and Number Processes

- *I can use addition, subtraction, multiplication and division when solving problems, making best use of the mental strategies and written skills I have developed.* ***MNU 1-03a***

Expressions and Equations

- When a picture or symbol is used to replace a number in a number statement, I can find its value using my knowledge of number facts and explain my thinking to others. **MTH 1-15b**

## Resources

These should be available for pupils to use if they wish:

- blocks, counters or other manipulatives, e.g. toy animals or pictures of animals
- paper and pencil for drawing

## Before they start...

Pupils should be able to add and subtract within 20 and should understand the vocabulary 'more' and 'less'.

## Let's go!

The children are talking about the pets they own. There are four more goldfish than snakes. The total number of rabbits is one less than the total number of goldfish. Six of the animals are furry. There are two dogs. **How many of each type of pet are there?** Show your thinking.

## Enabling prompt

If pupils are unable to get started on the problem, or do not understand the problem, one of the following prompts may help:

- How might animals / blocks / counters help you?
- Would it help to draw something?

- Work out the furry animals first.
- How many furry animals are there? How many are dogs? How many must be rabbits?
- How does knowing this help us?
- Does knowing that there are 12 animals help?

## Extension activity

- Make up a problem like this for others to solve. Think about changing the number of animals. Make sure you have an answer for your problem.

## Let's discuss!

When pupils have had an opportunity to solve the problem, teachers should gather methods that they have used to record their thinking. The focus should be on the second part of the problem, 'Show your thinking'.

Teachers can facilitate discussion by using prompts such as:

- **What information got you started?**
- **How many different ways do we have of showing our thinking?**

Some pupils may use concrete materials or pictures to represent and solve this problem. They can be encouraged to demonstrate and share their thinking. Attention should be given to how they recorded the process rather than a solution alone.

Some pupils may use a diagram or bar model to get started. Once they have found this information the teacher can ask **How does knowing the number of rabbits help us?**

| 6 furry animals | |
|---|---|
| 2 dogs | 4 rabbits |

Some pupils may use numbers. Starting with the knowledge that there are 6 furry animals and 2 are dogs they may record 6 – 2 = 4 or 2 + 4 = 6. Teachers may wish to draw attention to the relationship between these calculations. From here, pupils may continue to use numbers to work through the information.

2 dogs + ? rabbits = 6 furry animals, so there are 4 rabbits

? goldfish – 1 = 4, so there are 5 goldfish

? snakes + 4 = 5 goldfish, so 1 snake

After the discussion, pupils may wish to explore further using the suggestions in the Extension activity.

## Let's check!

**Main problem:** 1 snake + 5 goldfish + 4 rabbits + 2 dogs = 12 pets in total

**Extension activity:** The problems created by the pupils can be explored as a class or made available for peers to attempt and give feedback on.

## Let's reflect!

When pupils have had an opportunity to discuss their methods and solutions, teachers may wish to facilitate a discussion using questions such as **What helped you to understand the problem? How did you know your answer made sense?** This problem can be used to explore the idea of justifying strategies, with teachers asking **Why did you use the strategy you did?** Exploring the Extension Prompt can support pupils in giving each other feedback using agreed criteria.

## Teacher Guide Reference

- Early Level Teacher Guide Unit 3
- First Level Teacher Guide Unit 3

# Strawberry Plants

## Experiences and Outcomes

Number and Number Processes

- *I can use addition, subtraction, multiplication and division when solving problems, making best use of the mental strategies and written skills I have developed.* ***MNU 1-03a***

## Resources

These should be available for pupils to use if they wish:

- scrap paper or whiteboards for jottings
- graduated or empty number lines

## Before they start...

Pupils should be able to add and subtract two-digit and single-digit whole numbers. It may also be helpful to check that pupils know that strawberries grow on strawberry plants.

## Let's go!

Finlay has nine strawberry plants.

Finlay wants to collect 20 strawberries to decorate a special dessert his dad is making.

Finlay's dad says that he must collect **all** the strawberries from one plant before starting to pick more strawberries from another plant.

**How many different ways can he collect exactly 20 strawberries?**

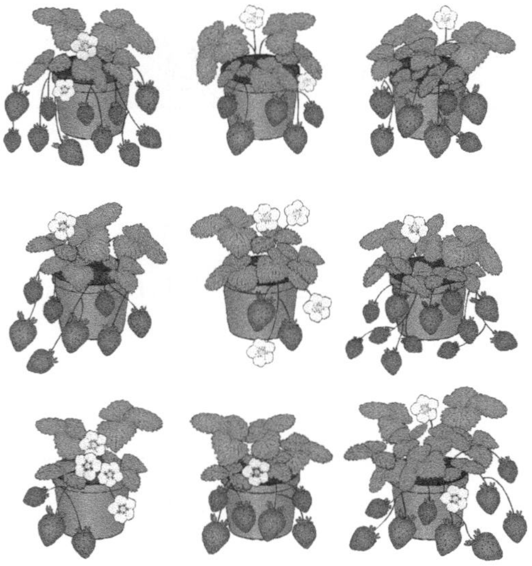

## Enabling prompt

If pupils are unable to get started on the problem, or do not understand the problem, one of the following prompts may help:

- Can you find pairs of numbers that total 20?
- Could Finlay collect 20 strawberries from the plants with 11, 5, 7 and 8 strawberries on them? Why/why not?
- Does it help if we order the numbers from smallest to largest? Could that help you find pairs that add to 20?

## Extension activity

- How can you make 20 with the smallest number of strawberry plants? How can you make 20 with the largest number of strawberry plants?
- How many desserts could Finlay decorate if he used all of the strawberries on his strawberry plants? How do you know?

### Let's discuss!

Giving pupils time to independently puzzle over the task, try different strategies in order to begin the task, make errors and self-correct is important. Teachers may wish to gather pupils together *briefly* after approximately 10 minutes to share strategies / methods that are being used. Teachers may ask **What have you noticed? What have you tried? How are you recording your answers?** Allow time for pupils to share how they approached the task, what they have tried and if there are patterns that were spotted. Teachers may wish to ask **Did you find all of the ways to make 20? How do you know?**

### Let's check!

**Main problem:**

| | | | | | |
|---|---|---|---|---|---|
| 12, 8 | 12, 6, 2 | 12, 5, 3 | 11, 9 | 11, 7, 2 | 11, 6, 3 |
| 9, 8, 3 | 9, 6, 2, 3 | 9, 6, 5 | 8, 7, 5 | 7, 6, 5, 2 | |

**Extension activity:**

There are 63 strawberries altogether, enough for 3 desserts.

## Let's reflect!

When pupils have had an opportunity to discuss their methods and solutions, teachers may wish to facilitate discussion around the information that pupils focused on in the initial stages of the problem. They may ask **How do you feel when you read a problem like this? Where did you decide to begin the task? What numbers did you use first? Did spotting numbers that add up to 10 help? What method did you use to find numbers that add up to 20?** The acknowledgement that problems can feel overwhelming is important. Discussion can then follow as to how pupils move on from that situation. It may also be beneficial to discuss how pupils became 'unstuck'.

## Teacher Guide Reference

- First Level Teacher Guide Unit 3

# Sweeties in Jars

## Experiences and Outcomes

Number and Number Processes

- *I can use addition, subtraction, multiplication and division when solving problems, making best use of the mental strategies and written skills I have developed.* ***MNU 1-03a***

## Resources

These should be available for pupils to use if they wish:

- pencil and paper
- number lines or hundred squares
- Resource sheet: Sweeties in a Jar – Empty Jars
- concrete materials to model the sweeties

## Before they start...

Pupils should be able to add one-digit whole numbers within 16 and should know what odd numbers are.

## Let's go!

You have four jars and 16 sweeties. Put the sweeties in the jars so that each jar has an odd number of sweeties. **How many different combinations can you make?**

## Enabling prompt

- What if you have eight sweeties and four jars? Share the sweeties between the jars so that each jar has an odd number. What could you use to help you?
- What if the first jar has five sweeties and the second jar has five sweeties?

## Extension activity

- You have 30 apples. Place the apples so that each bowl has twice as many as the bowl before. **Is there more than one way to do this?**
- You have 18 apples. Place the apples so that each bowl has one more than the one before it. **How many different ways can you do this? Can you still do this with only 16 apples?** (Note for the teacher: the second question requires working with halves.)

## Let's discuss!

Giving pupils time to independently puzzle over the task, try different strategies in order to begin the task, make errors and self-correct is important. Teachers may wish to gather pupils together *briefly* after approximately 10 minutes to share the strategies or methods that are being used. Teachers may ask **What have you noticed? What have you tried? How are you recording your answers?** Allow time for pupils to share how they approached the task, what they have tried and if there were patterns that were spotted. It may also be beneficial to discuss the meaning of 'odd' in relation to numbers.

This is a problem that may be tackled either independently or collaboratively.

Some pupils may start by writing random odd numbers that total 16. If this is the case, it may be helpful to ask the question **How do you know you have all of the answers?** to try to prompt pupils to either work more systematically or alter how they have recorded their answers.

Others may work systematically by beginning with 1 sweetie in the first jar, 1 sweetie in the second jar and then making combinations of 14, before moving on to 1 sweetie in the first jar, 3 sweeties in the second jar and so on.

Problem solving can be haphazard and pupils should be allowed to find out for themselves that there is a need to record thinking. If pupils are becoming frustrated that they cannot recall which consecutive numbers they have used or which numbers they have created, the teacher can ask them how they can keep a record of what they did. Encouraging pupils to use the recording sheet may help them with recording their thinking. There may be some discussion around whether 1, 1, 7 and 7 is the same as 7, 1, 7 and 1 and the replication of answers. Teachers can gather a great deal of information at this time by observing methods and conversations. Attention can also be drawn to the advantages of having a way of recording answers that allows the pupil to know they have explored all possible options.

When pupils have had an opportunity to solve the problem, teachers should gather methods that pupils have used to record their thinking. Teachers may then facilitate discussion by using prompts such as:

- **How did you start this problem?**
- **How did you ensure you were able to calculate all the answers? Is 1, 13, 1, 1 the same as 13, 1, 1, 1?**
- **How did you record your answers so that you didn't lose track?**

## Let's check!

**Main problem:**

There are nine solutions to the main problem:

| | | | |
|---|---|---|---|
| 1 | 1 | 1 | 13 |
| 1 | 1 | 3 | 11 |
| 1 | 1 | 5 | 9 |
| 1 | 1 | 7 | 7 |
| 1 | 3 | 3 | 9 |
| 1 | 3 | 5 | 7 |
| 1 | 5 | 5 | 5 |
| 3 | 3 | 3 | 7 |
| 3 | 3 | 5 | 5 |

**Extension activity**:

- 18 apples can be organised into 3, 4, 5 and 6 apples in the jars.
  16 apples can be organised as 2½, 3½, 4½ and 5½ apples in the bowls.
- 30 apples can be organised into 2, 4, 8 and 16 apples in the bowls.

## Let's reflect!

After pupils have had the opportunity to discuss their methods and solutions, teachers may wish to facilitate discussion using questions such as **How did you feel when you first read the problem? Did your feelings change as the time progressed?** The advantages of working systematically should be discussed with reference to how this helps us share our thinking and learn new strategies. This is particularly important where finding all the possible answers is necessary. If pupils collaborated on this problem, ask **What were the advantages or disadvantages of working with others on this problem?**

## Teacher Guide Reference

- Early Level Teacher Guide Unit 2
- First Level Teacher Guide Unit 3

# The Art Lesson

## Experiences and Outcomes

Number and Number Processes

- *I can use addition, subtraction, multiplication and division when solving problems, making best use of the mental strategies and written skills I have developed.* ***MNU 1-03a***

## Resources

These should be available for pupils to use if they wish:

- blocks
- Resource sheet: The Art Lesson Blank Grid

## Before they start...

Pupils should be able to add and subtract within 30.

## Let's go!

The class are enjoying an art lesson. The teacher has marked out a space on the wall like this so that all 24 children in the class can display their work:

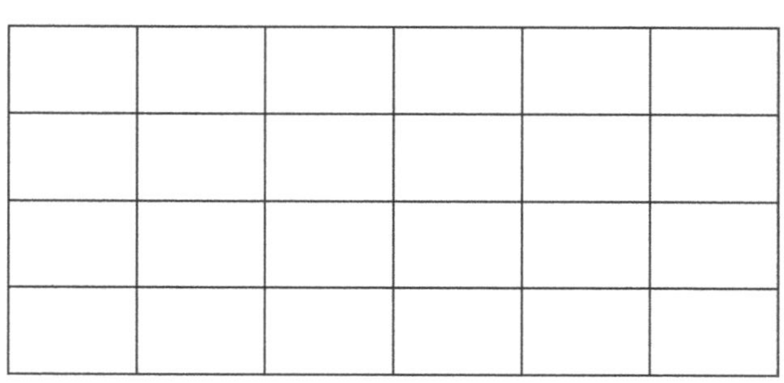

As each child finishes, the teacher puts their picture in one of the spaces. Nuria realises that there will be six empty spaces as six children are away at a music lesson.

At the end of the lesson Finlay says to Isla, 'Do you think our teacher meant to display the pictures like that?'

'What do you mean?' asks Isla.

'Well, there are an even number of pictures in every row and every column', Finlay replies.

**How might the teacher have arranged the pictures?**

## Enabling prompt

If pupils are unable to get started on the problem, or do not understand the problem, issue them with a 3 × 3 square and ask them to arrange six pictures so there are an even number in each row and each column.

## Extension activity

- There are over 72 different ways to do this. **How many different ways can you find?**

## Let's discuss!

This problem requires perseverance to get started. Many pupils will start with a trial-and-error approach. When pupils have found one solution, it is possible to build upon this. Working in collaboration may support finding an initial solution. Teachers can gather a great deal of information at this time by observing methods and conversations.

It may be worth gathering the class after a short time and recording what they have noticed.

Teachers may facilitate discussion by using prompts such as:

- **What do you notice about the empty spaces in each row and column?**
- **Why must there be pictures in all the rows?**
- **Why can there not be four full columns?**
- **What even numbers will help us with this problem?**

Pupils may then continue to explore solutions.

Some pupils may use the grid provided and place blocks to represent pictures. This allows them to make adjustments with ease. Teachers can ask **When you found a solution, how did it help you to find others? In what ways are our solutions the same?**

Teachers may wish to draw attention to solutions that are reflections or rotations of each other. This offers an opportunity to discuss using what is known to find other solutions.

After the discussion, pupils may wish to gather the various solutions they have found and continue to add to them over time. There are multiple solutions to be discovered.

## Let's check!

**Main problem:** There are numerous solutions. Two are given here. They can be used to look for similarities. Both have a row of 6 pictures and 3 rows of 4.

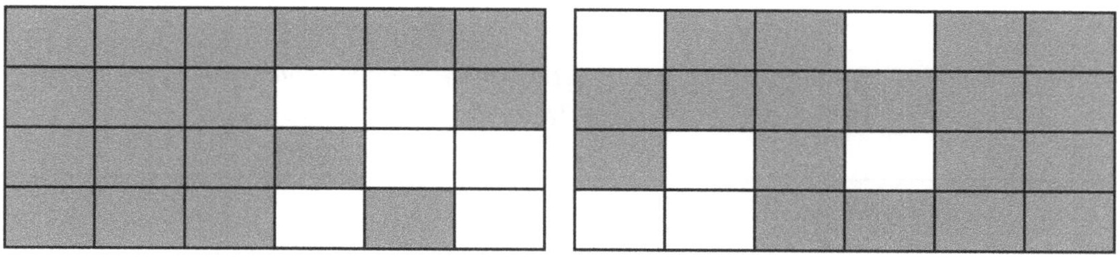

## Let's reflect!

When pupils have had an opportunity to discuss their methods and solutions, teachers can facilitate discussion using questions such as **How did you feel when you first saw the problem? Did your feelings change as you worked through the problem?** Some pupils may find the perseverance required to find solutions to be a barrier. Teachers may ask **Why is it useful to use the ideas of others? What advantages are there to sharing your thinking with another person?** Teachers can encourage pupils to share strategies that they use to keep going.

## Teacher Guide Reference

- First Level Teacher Guide Unit 3

# The Netball Team

## Experiences and Outcomes

Number and Number Processes

- *I can use addition, subtraction, multiplication and division when solving problems, making best use of the mental strategies and written skills I have developed.* ***MNU 1-03a***

## Resources

These should be available for pupils to use if they wish:

- paper and pencil

## Before they start...

Pupils should be able to add and subtract two-digit numbers.

## Let's go!

The school is buying sports shirts for the netball team to wear in a competition. The shirts have the children's names on the back. Finlay's shirt will cost £22. Isla's shirt costs £18. Nuria and Amman's shirts will cost £20 each.

Each shirt costs the same but there is a charge for each of the letters on the back.

**How much does a shirt without a name on it cost? How much would a shirt with your name on it cost?**

## Enabling prompt

If pupils are unable to get started on the problem, or do not understand the problem, one of the following prompts may help:

- How many letters are in each child's name?
- How many more letters are in Nuria's name than Isla's name? How much more expensive is Nuria's shirt than Isla's shirt?

## Extension activity

- Create your own charge per letter, and for the shirt, and challenge a partner to solve your puzzle.

## Let's discuss!

Teachers may need to ensure that pupils understand the problem before allowing them time to explore their thinking. It is a suitable context for discussing sales with basic features and add-ons available for a price.

When pupils have had an opportunity to solve the problem, teachers should gather methods that pupils have used to record their thinking. Teachers may then facilitate discussion by using prompts such as:

- **What did you notice about the difference between the price of each child's shirt?**
- **How did you decide what was important information?**
- **How did you keep track of your calculations?**

Some pupils may use algebraic thinking and record their thinking in a way such as:

 + 6 letters = £22

 + 4 letters = £18

 + 5 letters = £20

By looking at the difference they then know that each letter increases the price by £2. Then they can work out the cost of lettering for each shirt and subtract from the price.

Some pupils may use a guess and check method. It is worth listening to the numbers pupils use to gauge their sense of number. As a shirt with five letters is £20 and there is a cost to the basic shirt, the number per letter could not be greater than £3. The discussion around how pupils select starting points can be worth unpicking. The teacher may ask **How does your first try help you decide what to do next?** Pupils should be able to articulate that their first attempt directs them to try larger or smaller numbers.

After the discussion, pupils may wish to explore further using the suggestions in the Extension activity.

## Let's check!

**Main problem:** A shirt costs £10. The child's own name will be £10 + £2 per letter in their name.

**Extension activity:** Solutions will depend upon pupils' own problems. Pupils should ensure that they have checked that their solution to their own problem is correct.

## Let's reflect!

When pupils have had an opportunity to discuss their methods and solutions, teachers may wish to facilitate discussion using questions such as **What helped you to understand the problem? Was there any point at which you felt frustrated?** Exploring the Extension activity can support pupils in being creative in their thinking.

## Teacher Guide Reference

- First Level Teacher Guide Unit 3

# The Water Challenge

## Experiences and Outcomes

Number and Number Processes

- *I can use addition, subtraction, multiplication and division when solving problems, making best use of the mental strategies and written skills I have developed.* ***MNU 1-03a***

## Resources

These should be available for pupils to use if they wish:

- a variety of empty containers such as yogurt pots and tins (larger ones labelled 7 and smaller ones labelled 3) – these do not need to be proportionally exact
- empty bucket or washing up bowl
- water

## Before they start...

Pupils should be able to add and subtract within 10.

## Let's go!

The class have the water tray out in the classroom. Nuria has a container that holds 7 litres of water and Finlay has a container that holds 3 litres of water.

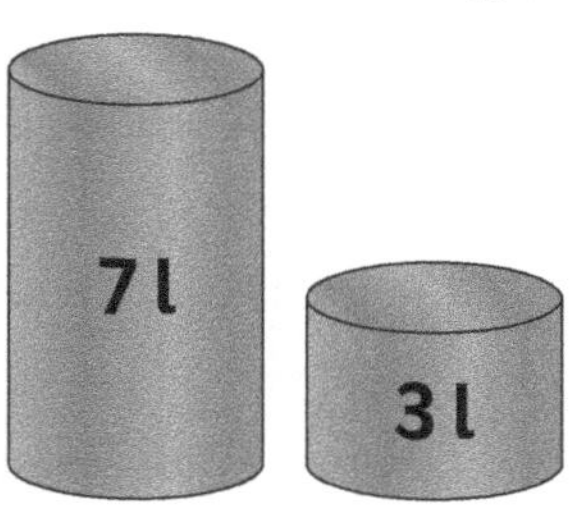


'I wish they had markings to show the measurements', said Amman.

'We can measure out every amount from 1 litre to 10 litres with just those two containers', said Isla. 'I have a big empty bucket we can use to help us.'

Is Isla correct?

## Enabling prompt

If pupils are unable to get started on the problem, or do not understand the problem, one of the following prompts may help:

- How much does each container hold? Use two of the labelled containers to measure out 10 litres.
- How can we use what we know to measure out 4 litres?
- Try in this order, 10l, 6l, 9l 4l,1l, 2l, 8l, 5l.
- Keep a record of the moves you are making.
- Remember that you can use the empty container to help.

## Extension activity

- Try to make each amount from 1 litre to 10 litres using two containers that hold even amounts of water rather than odd.
- Try to make each amount from 1 litre to 10 litres using two containers that hold the same amount of water.

## Let's discuss!

Teachers can give pupils the containers and discuss what they notice. Pupils may notice that some are labelled and some are not. The teacher should share the problem with the pupils and they should be given time to digest the problem for themselves before starting. If pupils are finding it challenging to understand what the question is asking, then teachers may ask **How can you make 10 litres? How can you find 4 litres?**

This problem requires a systematic method of recording what is happening, as it is easy to lose the sequence that pupils have discovered. Problem solving can be haphazard and pupils should be allowed to find out for themselves that there is a need to record their thinking. If pupils are becoming frustrated because they cannot recall how they solved the problem, the teacher should ask them how they can keep a record of what they do. Teachers can gather a great deal of information at this time by observing methods and conversations.

When pupils have had an opportunity to solve the problem, teachers should gather methods that pupils have used to record their thinking. Teachers may then facilitate discussion by using prompts such as:

- **What advice would you give to another class doing this problem?**
- **Why were some amounts more straightforward to find?**
- **What did you find out?**

Some pupils may use the labelled containers and empty container to find out which amounts can be created in a straightforward way and record those first. Having worked out the amounts that can be made in a few moves, pupils will then keep exploring to see which other amounts they can find. When discussing these, teachers may ask questions such as **If you can make 3 litres, what other amounts can you make using 3 litres?**

The recording of solutions can be compared and discussed. A record such as the one below can be used to ask **What does this image show us? What helps us to understand it?**

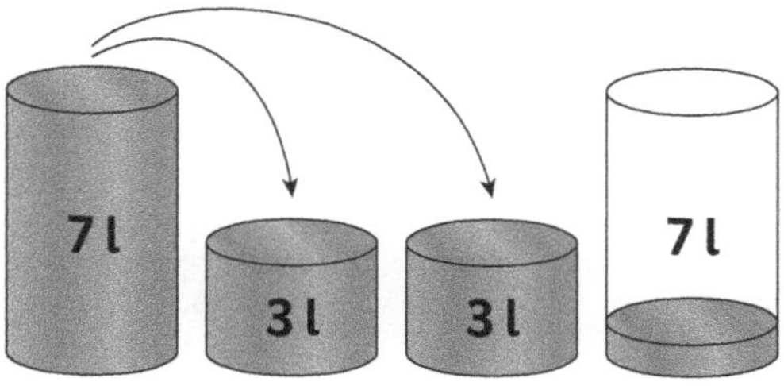


After the discussion, pupils may wish to explore further using the suggestions in the Extension activity.

## Let's check!

**Main problem:** Isla is correct.

1. Fill the 7l container and pour it into the 3l container. Empty the 3l container and refill from the 7l. There is now 1l left in the 7l container.
2. Fill the 3l container and pour it into the 7l container. Repeat this. Fill the 3l container again and pour it into the 7l container until the 7l container is full. There will be 2l left in the 3l container.
3. Fill the 3l container.
4. Fill the 7l container and pour it into the 3l container until full. 4l will be left.
5. Pour the 7l container into the empty bucket. Repeat. There is 14l in the bucket. Take 3l out of this which leaves 11l, do this again and it leaves 8l. Take 3l out of this and there is 5l.
6. Fill the 3l container and pour into an empty container. Do this twice.
7. Fill the 7l container.
8. Pour the 7l container into the empty bucket. Repeat. There is 14l in the bucket. Take 3l out of this, which leaves 11l. Do this again and it leaves 8l.
9. Fill the 3l container and pour into the empty container. Do this three times.
10. Fill the 3l container and 7l containers and pour into the empty container.

Pupils may have other solutions.

## Let's reflect!

When pupils have had an opportunity to discuss their methods and solutions, teachers may wish to facilitate discussion using questions such as **How did you feel when you first saw the problem? Did your feelings change as you worked through the problem? Did you use any number facts as you worked through the problem?** Some pupils may find the perseverance required to find solutions to be a barrier. Teachers can encourage pupils to share strategies that they use to keep going. They may ask **Why is it helpful to be able to quickly add and subtract for this problem?**

## Teacher Guide Reference

- First Level Teacher Guide Unit 3

# 100 Maze

## Experiences and Outcomes

Number and Number Processes

- *I can use addition, subtraction, multiplication and division when solving problems, making best use of the mental strategies and written skills I have developed.* ***MNU 1-03a***

## Resources

These should be available for pupils to use if they wish:

- Resource sheet: 100 Maze
- paper and pencils for jottings

## Before they start...

Pupils should be able to count on to keep a running total.

## Let's go!

Find a way to move through this maze from the START to the FINISH so that the numbers you pass through total to exactly 100. You are only allowed to go through a square once. You may move vertically or horizontally but not diagonally. The thick black lines are barriers that cannot be crossed. Show your route clearly on the maze.

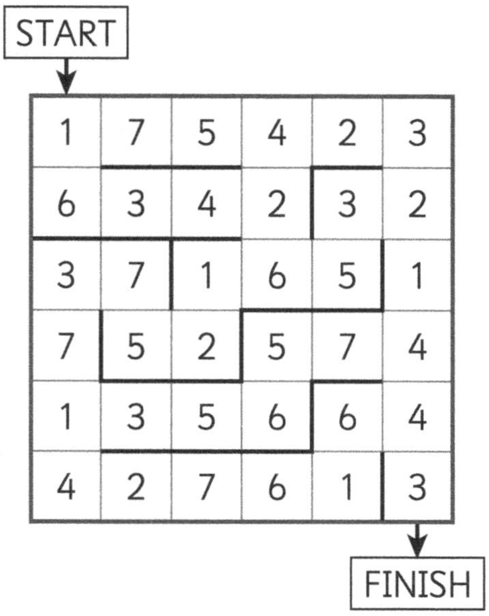

## Enabling prompt

Pupils who are struggling to make a start on the 100 Maze may be given this task.

In these grids you can only move left, right, up or down. Diagonal moves are not allowed. Find the total of the numbers in the squares that the grey line passes through in each of the grids.

**Grid 1**

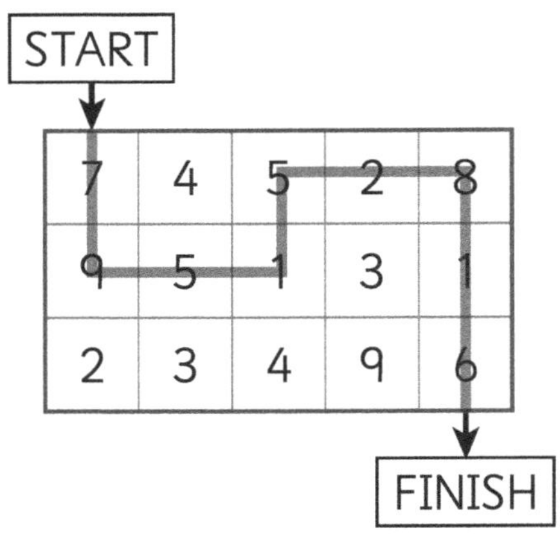


**Grid 2**

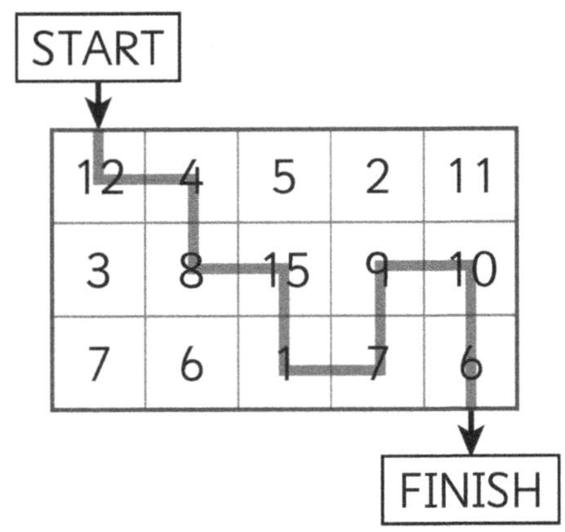


## Extension activity

How many different totals do you think there are altogether for the routes through the maze? What do you think the lowest total is? What do you think the highest total is?

### Let's discuss!

It is likely that pupils will begin by randomly plotting routes through the maze and totaling the numbers. Some may then try to adjust their route to add in higher numbers or lower numbers depending on whether their total was more than or less than 100. This, or other strategic approaches, should be shared at appropriate points. Pupils should be encouraged to share their approaches to recording their routes and checking that they were not repeating the same route. Some discussion of the desired total of 100 and the size of the numbers within the cells would be worthwhile. The teacher might ask **Roughly how many cells do you think your route might need to go through to get close to 100? What makes you think that?**

## Let's check!

**Enabling prompt:**

GRID 1 total: 44

GRID 2 total: 72

**Main problem:**

There may be several correct solutions offered, for example:

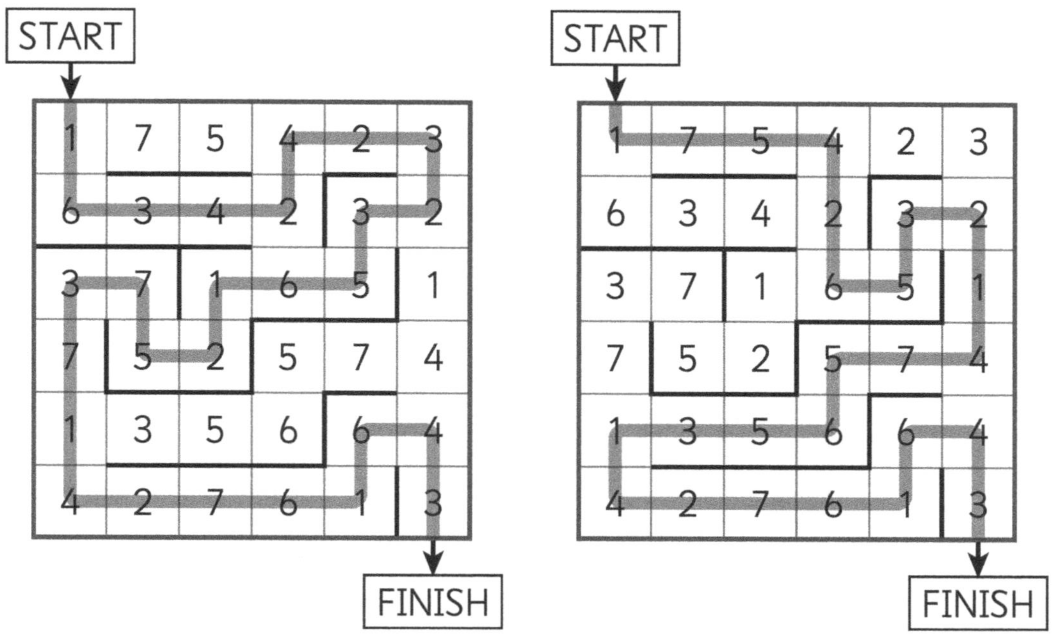


**Extension activity:**

Pupils' own answers

## Let's reflect!

In this problem, pupils may have experienced some disappointment at being close to the desired total only to have to go back to the beginning when they were not able to adjust their route to make 100. The teacher may ask **How did you feel when you got very close to the total? Did anyone have to go back to the beginning when a route did not make 100? How did this make you feel? Were you able to adjust a route that was close to 100 by looking at some of the numbers you had not used? When you did eventually make 100 after adjusting your route, how did that feel?**

## Teacher Guide Reference

- First Level Teacher Guide Unit 3

# Functions

## Experiences and Outcomes

Number and Number Processes

- *I can use addition, subtraction, multiplication and division when solving problems, making best use of the mental strategies and written skills I have developed.* ***MNU 1-03a***

## Resources

These should be available for pupils to use if they wish:

- Resource sheet: Functions
- paper and pencil for jottings

## Before they start...

Pupils should be confident in adding, subtracting, multiplying and dividing single-digit and two-digit numbers.

## Let's go!

This is a function machine. When a number is fed into the machine, a calculation is carried out at each step and the answer comes out at the end.

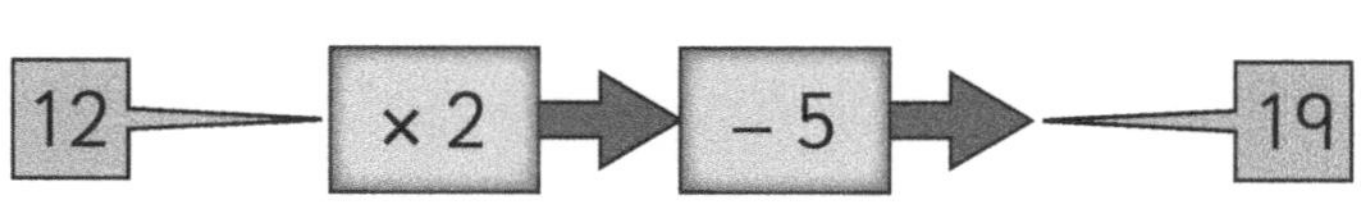


This is a different function machine. When 6 is fed in, 26 comes out. When 3 is fed in, 17 comes out and when 10 is fed in, 38 comes out. Unfortunately, the calculations that it does have been covered over with paint splodges. **What do you think they are?**

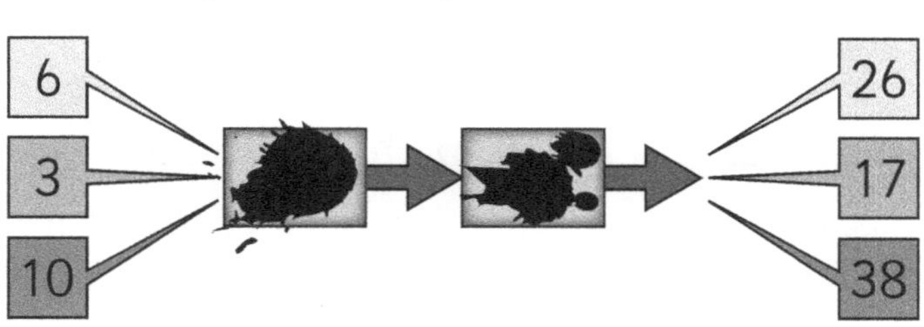

The operator of this function machine has a list of numbers but can't remember which ones went into the function machine and which ones came out. Sort these numbers into pairs for the operator, making it clear what the input number is and what the output number is in each pair.

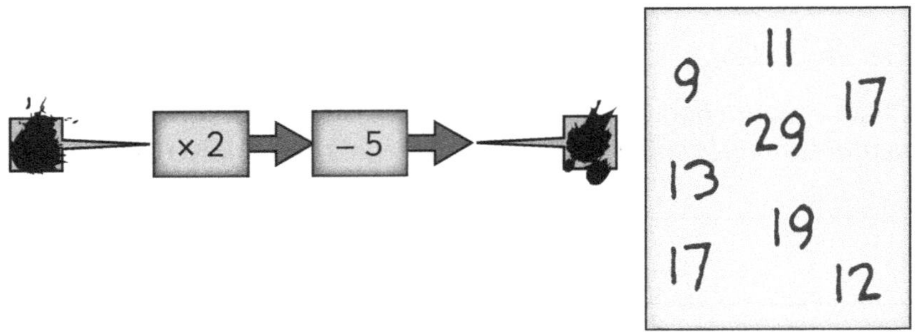


## Enabling prompt

Pupils who struggle to make a start to the paint splodge problem may be asked to choose from these pairs of functions, trying each one to see which they think fits.

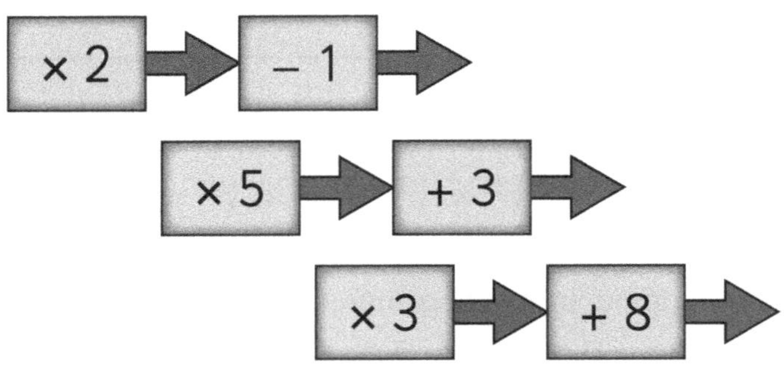


## Extension activity

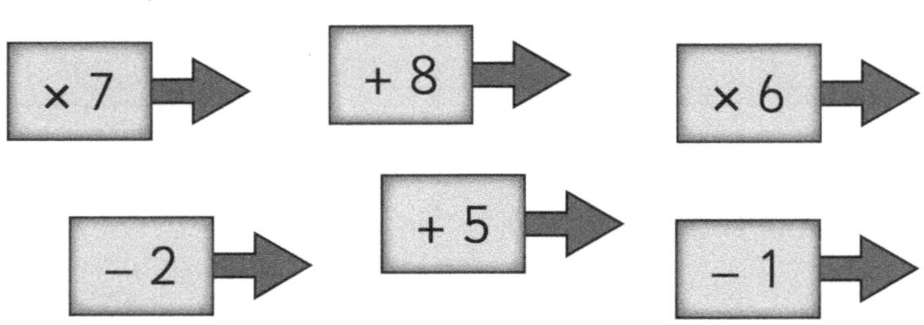


A function machine operator needs to make a machine that will give the number 160 when 23 is fed in and 104 when 15 is fed in. The operator has these functions to choose from. Which ones do you think he should choose for the machine?

This is a 3-step function machine.

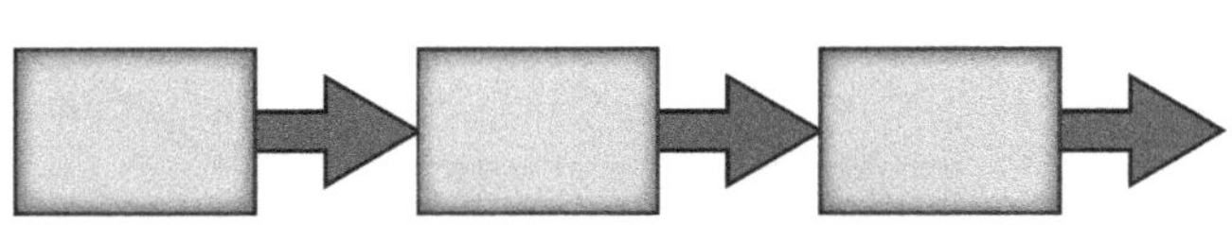

What might the functions be for Input 4 and Output 38 in this 3-step function machine? Does this work for Input 10 and Output 74? Can you create a 3-step function machine that works for both of these?

## Let's discuss!

The teacher should lead an initial discussion of the example given at the beginning to ensure that pupils are clear about the context of the function machine. Pupils should then be left to try different functions for the initial activity. When something that works for one pair of numbers is found, the teacher may say **Try your functions for the next pair to see if it works for them too**. For the second activity, involving sorting numbers into pairs, pupils may notice that 17 appears twice and deduce that it must be both an 'Input' number and an 'Output' number.

In the Extension activity, pupils may ask if they can use more than two of the function machines, or to add in a function of their own choosing. They should be given the freedom to use their creativity, even although ultimately it might result in a relatively simple solution being presented.

## Let's check!

**Main problem:**

The functions that have been hidden by paint are '× 3', '+ 8'.

The number pairs are: In 9 > Out 13; In 11 > Out 17; In 12 > Out 19; In 17 > Out 29

**Extension activity:**

'× 7' then '– 1' will give the correct 'Output' numbers here.

There are numerous possibilities for Input 4 and Output 38, such as + 5 × 4 + 2; –1 × 6 + 20. Similarly, for both Input 4 and Output 30 **and** Input 10 and Output 74 there are many possibilities, such as × 3 + 7 × 2 or × 2 × 3 + 14 or × 6 + 5 –1.

## Let's reflect!

Pupils should think about how they needed to persevere when trying various things as they worked towards solutions in this problem. The teacher may ask **How did it feel when you found something that worked for one pair of numbers but then discovered that it did not work for all of the pairs?** Pupils should also be encouraged to reflect on how incorrect ideas helped them to work towards solutions.

## Teacher Guide Reference

- First Level Teacher Guide Units 7 and 12

# Maths Mobiles

## Experiences and Outcomes

Number and Number Processes

- *I can use addition, subtraction, multiplication and division when solving problems, making best use of the mental strategies and written skills I have developed.* ***MNU 1-03a***

## Resources

These should be available for pupils to use if they wish:

- Resource sheet: Maths mobiles

## Before they start...

Pupils should be confident with addition and subtraction of single-digit and two-digit whole numbers.

## Let's go!

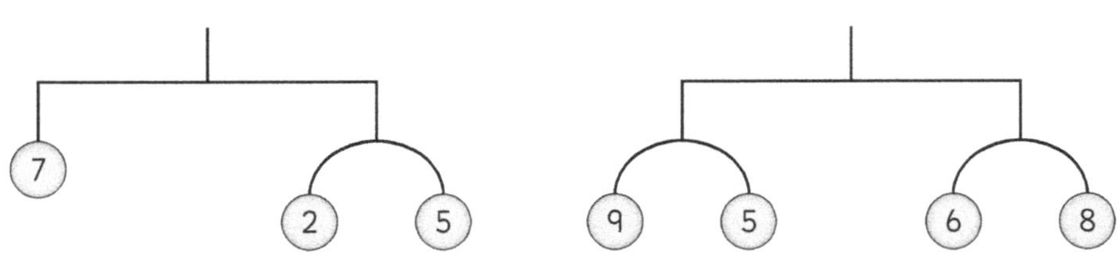


The children are making maths mobiles to hang up in class. To make their mobiles balance, they must make sure that the numbers on each arm have the same total.

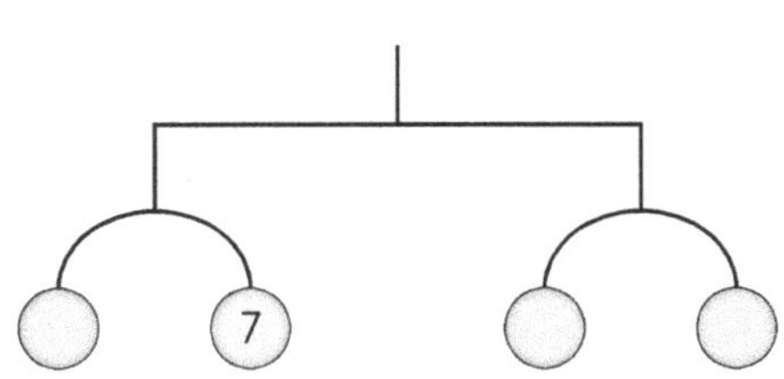


The children begin by completing this mobile.

When the mobiles are complete, Isla notices that there is only one 7 on each of their mobiles but that each mobile is different. How do you think the four children completed their mobiles?

Finlay makes this mobile and asks the others to add numbers to complete it.

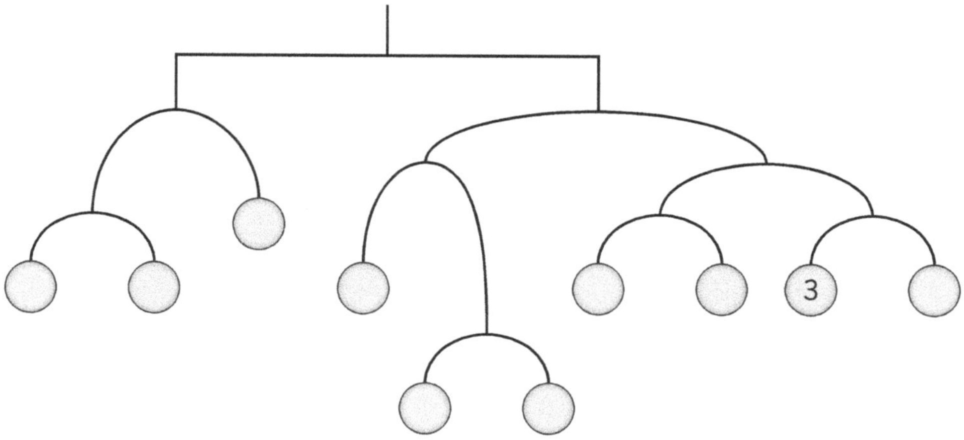


Find two different ways to do this.

## Enabling prompt

Pupils who are struggling to make a start to this problem may be given this version. The teacher should prompt with questions like What does this side make when we add the numbers? (pointing to the side with a 3 and a 7 added) Can you find another pair of numbers that add together to make 10 and put these into the mobile?

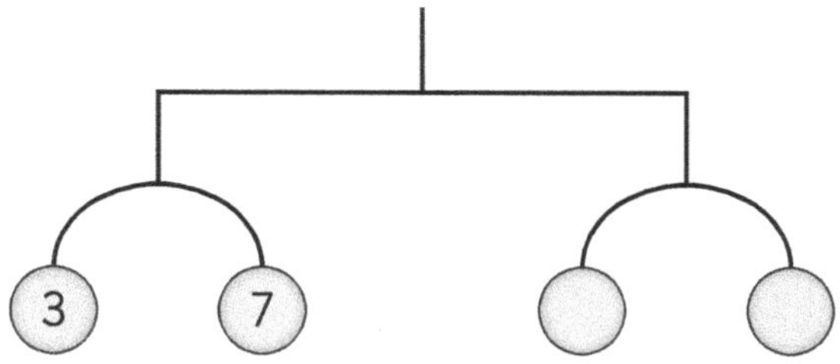


These pupils can then move onto this, or similar, with the teacher prompting as required.

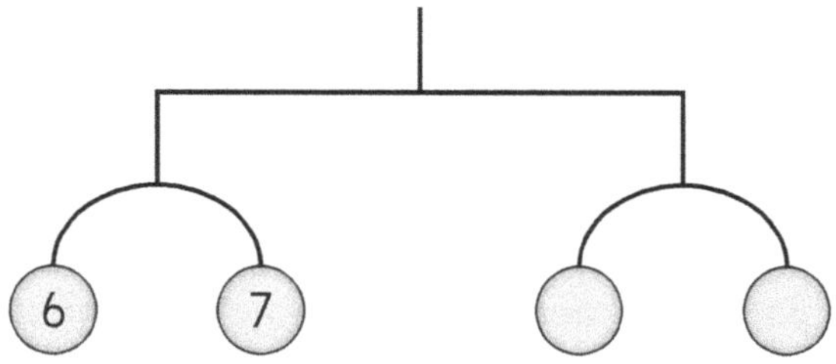

## Extension activity

In this balance mobile, all of the straight arms must balance. Complete this mobile.

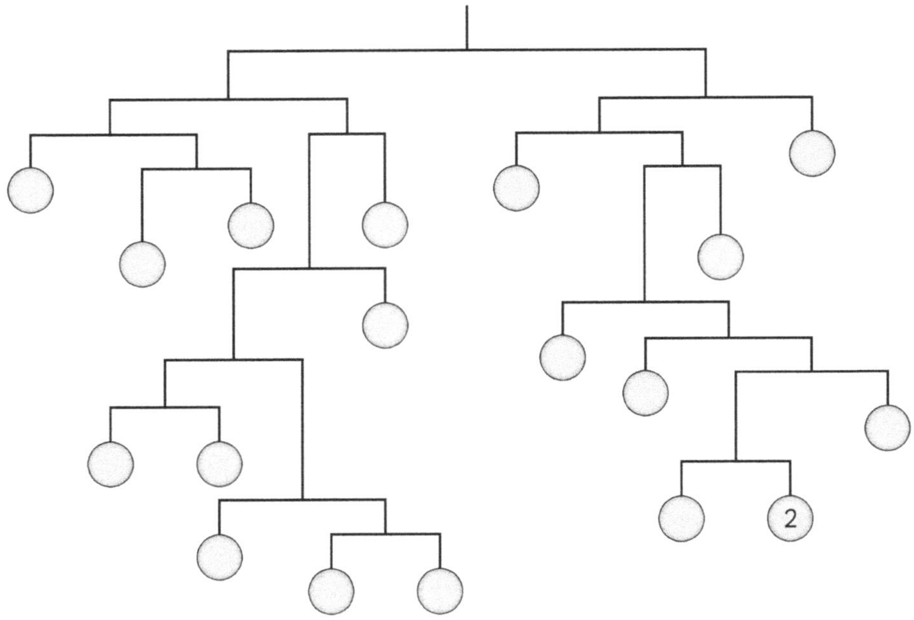


## Let's discuss!

Pupils should be given an opportunity to discuss and explain the balancing in the two examples that are given at the beginning of the problem. It is likely that pupils will notice that using known number facts will help them when working out four different ways to complete the first mobile. For example, when 3 is placed on the same arm as the 7, the total is 10. This means that 1 and 9 or 2 and 8 or 4 and 6 or 5 and 5 can be added to the other arm. There should be discussion of why 3 and 7 will not appear on this arm. Some pupils may choose to use larger numbers. For example, if 19 is placed on the same arm as the 7, then two numbers totalling 26 must appear on the other arm. This may lead to consideration and subsequent rejection of 17 and 9.

In Finlay's mobile, pupils may simply write numbers in each of the discs on the right-hand arm of the mobile. There should be discussion of how they can then ensure that the left-hand arm balances. The teacher may ask **How did you track your total on one arm and how did you use this to help you complete the second arm?**

## Let's check!

**Main problem:**

There are many possible solutions for the first mobile (with 7 already added) and for Finlay's mobile.

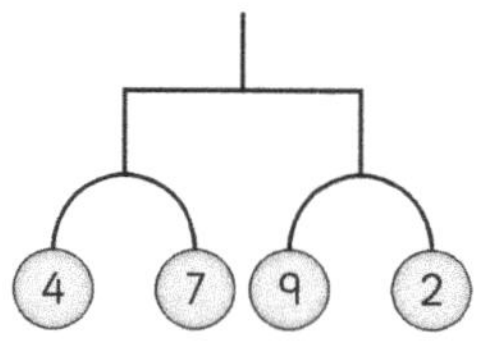


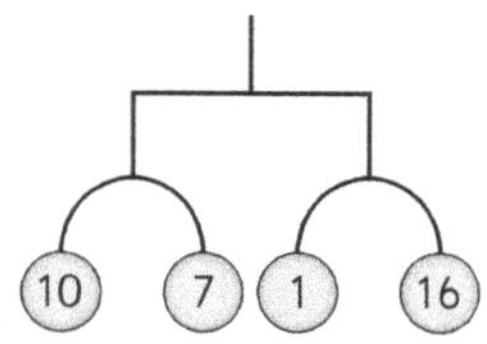


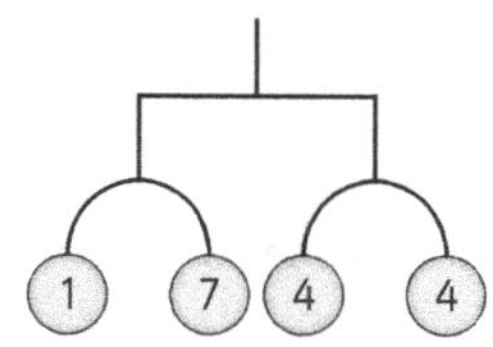


And so on.

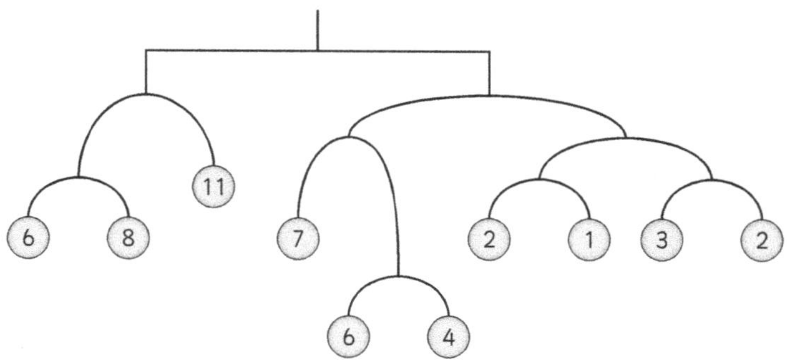


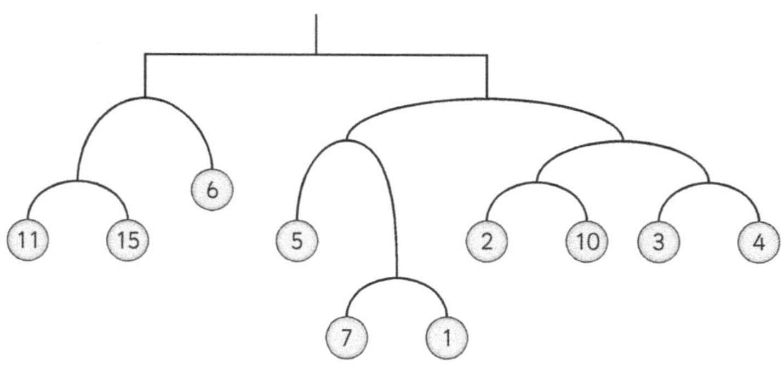


And so on.

**Extension activity:**

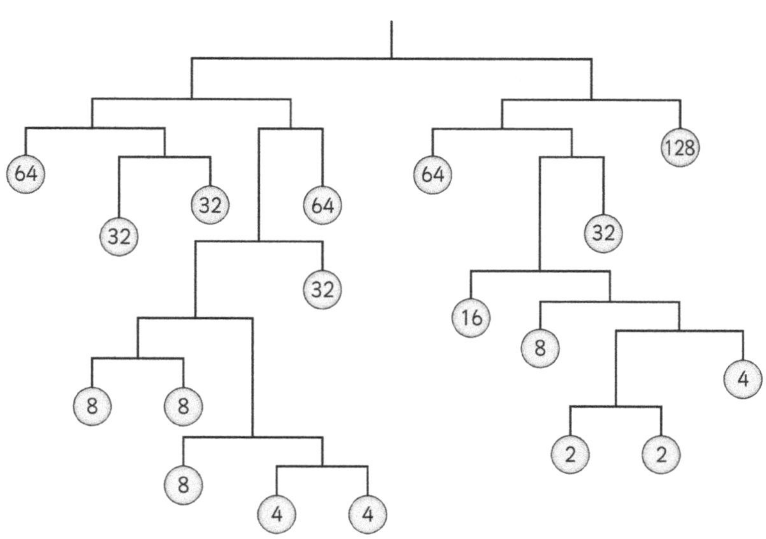

## Let's reflect!

Pupils should be encouraged to reflect on their strategies for simplifying this problem and how the use of a 'step-by-step' approach supported them. The teacher might say **Tell us how you organised your thinking and your working when you were solving these problems.** Pupils should be encouraged to think about how they felt when they realised that as the totals accumulated, they were dealing with larger numbers and had to rely more and more on known number facts. The teacher could ask **What did you feel like when you realised that you were dealing with larger numbers as you worked through these problems? How did it help you to recall number facts and use strategies for addition and subtraction that you learned previously?**

## Teacher Guide Reference

- First Level Teacher Guide Unit 3

# Isla's Party

## Experiences and Outcomes

Number and Number Processes

*Having explored fractions by taking part in practical activities, I can show my understanding of:*

- *how a single item can be shared equally*
- *the notation and vocabulary associated with fractions*
- *where simple fractions lie on the number line.* ***MNU 1-07a***

## Resources

These should be available for pupils to use if they wish:

- pencil and paper or whiteboards for informal jottings
- Resource sheet: Isla's Party

## Before they start...

Pupils should understand that quarters are created by splitting a whole into four equal parts.

### Let's go!

Isla is having a party and wants to order giant cookies for herself and her 17 friends. Each person should get ¼ of a cookie.

How many giant cookies should Isla order? Will there be any cookies left over? Draw a picture or a diagram to prove your answer.

## Enabling prompt

- Imagine this shape is a giant cookie. Can you split it into quarters? If everyone gets a quarter of a cookie, how many people can share this cookie?
- How many people can share two giant cookies?

## Extension activity

- If every child at the party gets ⅙ of a giant cookie, how many cookies are needed now? What do you notice about the size of the parts compared to the size of the parts when you split the cookie into quarters?
- Isla gives each guest a goody bag to take away with them after the party. If each guest gets ¼ of a bag of sweets, how many bags of sweets are needed?

## Let's discuss!

Giving pupils time to independently puzzle over the task, try different strategies in order to begin the task, make errors and self-correct is important. Teachers may wish to gather pupils together *briefly* after approximately 10 minutes to share strategies and methods that are being used. Teachers may ask **What have you noticed? What have you tried? How are you recording your answers?** Allow time for pupils to share how they approached the task, what they have tried and if there were patterns that were spotted.

Some pupils may start by drawing giant cookies and splitting these into quarters in order to work out how many cookies are needed for 18 children. If pupils have split the cookies into four but these are not four equal parts, teachers may wish to prompt by asking questions such as **Are these quarters or four pieces? What's the difference?**

Others may start by using their knowledge of four quarters making a whole to calculate that four children can share one giant cookie; therefore, eight children can share two giant cookies, twelve can share three giant cookies, etc.

Pupils should be encouraged to show *equal* parts in their drawings and diagrams to reinforce the concept that a quarter is one of four equal parts.

Teachers may wish to ask **I see you have used circles to show the cookies. What happens if you use a different shape? Do you still get the same answer?** in order to probe pupils' understanding of fractions. If pupils struggle with the concept that it doesn't matter if their cookie is a circle, rectangle or square and that the answer will be the same regardless of the shape, this shows that their conceptual understanding of fractions needs further work. This could possibly include more hands-on, practical experiences to allow them to fold, cut and split materials into 2, 3 and 4 equal parts. Fractions can be represented as part of a whole or unit, part of a set, a division of whole numbers or a number on a number line. Pupils also need plenty of experience working with 2D models (e.g. squares of paper), 3D models (e.g. playdoh), linear models (e.g. strip of paper) and discrete collections (e.g. a group of counters) to develop their conceptual understanding of fractions.

During discussions, teachers may wish to probe: **Does your answer change depending on the shape of the cookie, e.g. does it matter whether the cookie is circular or rectangular?**

When pupils have had an opportunity to solve the problem, teachers should gather methods that pupils have used to record their thinking. Teachers may then facilitate discussion by using prompts such as:

- **How did you start this problem?**
- **How did you decide to record your ideas?**
- **How did you check your answer?**
- **How much cookie is left over?** There are 2/4 of a cookie (or ½) left over.

**How many more people could Isla have given ¼ of a cookie to?** Isla could have invited 19 friends to her party as there was enough for 20 people.

## Let's check!

**Main problem:**

Isla needs to order five cookies.

**Extension activity**

- Three giant cookies are needed if every guest receives 1/6. Guests will get a smaller piece of cookie when they are given sixths compared to their earlier answer of quarters.
- Five bags

## Let's reflect!

After pupils have had the opportunity to discuss their methods and solutions, teachers may wish to facilitate discussion using questions such as **Where did you start? Did the first strategy you try work? What drawing or diagram worked best for you? Was there a particular shape that you were drawn towards to work out quarters? What was the trickiest part of this task for you?**

## Teacher Guide Reference

- First Level Teacher Guide Unit 5

# Block Letters

## Experiences and Outcomes

Number and Number Processes

- *I can estimate the area of a shape by counting squares or other methods.* ***MNU 1-11b***

## Resources

These should be available for pupils to use:

- pencil and paper or whiteboards for informal jottings
- 1 cm squared paper

## Before they start...

Pupils should know which letters are in their first name. Pupils should be able to count to 20.
Pupils should have some experience of counting squares and half squares to find the area of a shape.

## Let's go!

Finlay, Nuria, Isla and Amman are drawing their first names on 1-cm squared paper. Each letter is a block capital letter, 5 squares high and 5 squares wide like this:

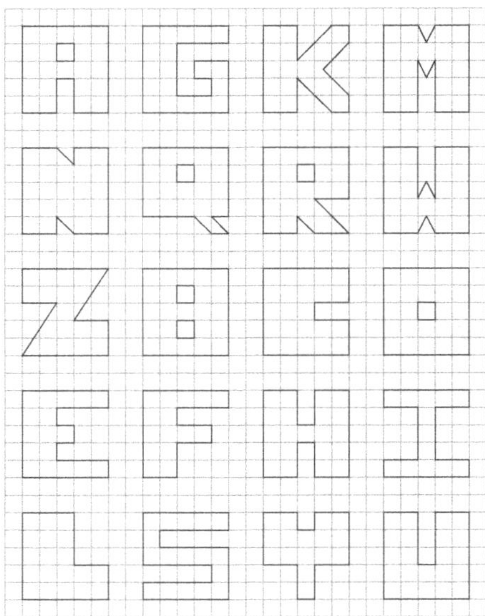

**Whose name do you think will have the greatest area?**
**Whose name do you think will have the smallest area?**
**Prove your predictions.**

## Enabling prompt

- Do you think there will be a link between the number of letters in the names and the area?
- What is the area of the first letter of Isla's name?

## Extension activity

- Which letters in the alphabet have the greatest area?
- Who in your class has a name that has a greater area than Finlay, Nuria, Isla or Amman?
- Estimate, then find the area of the name of someone in your family. Who do you think will have the name with the greatest area? Who will have the smallest area?

## Let's discuss!

Giving pupils time to independently puzzle over the task, try different strategies in order to begin the task, make errors and self-correct is important. Teachers may wish to gather pupils together *briefly* after approximately 10 minutes to discuss **What have you noticed? What have you tried? Are there any letters that you are struggling to write?** Allow time for pupils to share how they approached the task, what they have tried and if there are any letters that they are unable to create. It may be beneficial to point out that half squares may need to be used for some letters, as in the example letters.

This is a problem that may be tackled either independently or collaboratively.

Some pupils may begin to link the number of letters with the area of all the letters, hypothesising that the child with the longest name (Finlay) will have the name with the greatest area.

Others may start to explore the letters of the alphabet that have the greatest area, e.g. N, A, M, R or some of the smallest areas, e.g. I, L, Y, F, and link this to the children's names. Teachers may wish to ask **Why have you started with those letters? What are you finding?**

Problem solving can be haphazard and pupils should be allowed to find out for themselves that there is a need to record the area of each letter in the children's names. Teachers can gather a great deal of information at this time by observing methods and conversations. Asking **Were you surprised with your answer?** may elicit discussion around predictions that they may have made that may have been partially correct.

## Let's check!

**Main problem:**

Amman's name has the greatest area of 112 $cm^2$.

Finlay's name has an area of 107 $cm^2$.

Nuria's name has an area of 102.5 $cm^2$.

Isla's name has an area of 61 $cm^2$.

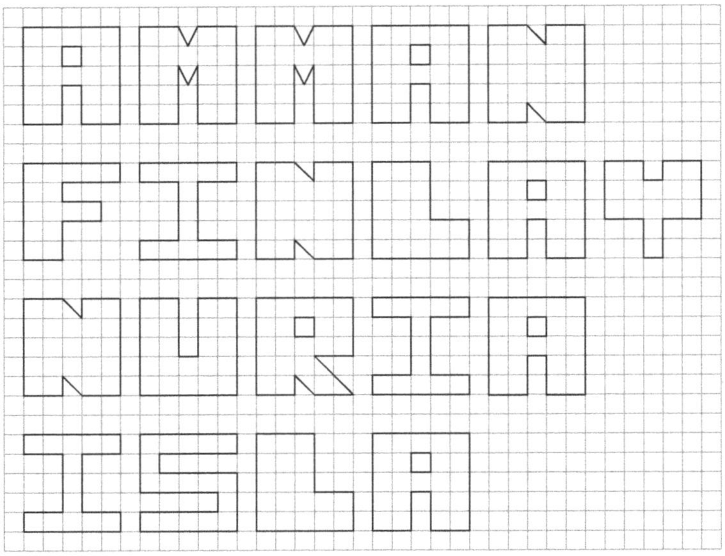

**Extension activity**

The letters D, O, N, B, C, U, M and A have the greatest area, so names that include five or more of these letters will have a greater area than Amman's 112 $cm^2$.

## Let's reflect!

This problem is ideal for allowing pupils to link mathematical concepts and transfer their knowledge from previous learning in mathematics. Due to the final answer not being the most obvious answer, teachers may wish to ask **Which name did you predict would have the largest area? How did you feel when you thought you were wrong?**

## Teacher Guide Reference

- Early Level Teacher Guide Unit 2
- First Level Teacher Guide Unit 8

# The Wonky Watch

## Experiences and Outcomes

Time

- *I can tell the time using 12-hour clocks, realising there is a link with 24-hour notation, explain how it impacts on my daily routine and ensure that I am organised and ready for events throughout my day.* ***MNU 1-10a***

## Resources

These should be available for pupils to use if they wish:

- clock faces
- Resource sheet: The Wonky Watch Blank Clock Faces

## Before they start...

Pupils should be familiar with reading analogue clocks. Pupils should also know that there are 60 minutes in an hour.

## Let's go!

Amman bought a watch for £1. Unfortunately, it gains 30 minutes every day! If Amman set his watch at noon on the day he bought it, how many days would it take for his watch to correctly show 12 o'clock again? Show your thinking.

## Enabling prompt

If pupils are unable to get started on the problem, or do not understand the problem, one of the following prompts may help:

- What time would the watch show after one day has passed? After two days?
- Use the blank clocks to help you keep a record of what the watch shows as each day passes.

## Extension activity

- If Amman's watch had a 24-hour digital display, how long would it take to show the correct time?

## Let's discuss!

When pupils have had an opportunity to solve the problem, teachers should gather methods that they have used to record their thinking. The focus should be on the second part of the problem, 'Show your thinking'.

Teachers can facilitate discussion by using prompts such as

- **How did your knowledge of telling the time help you to start this problem?**
- **How did you organise your thoughts?**

Some pupils may record their solution using an organised list or table. Working systematically can support pupils to feel confident about their final answer. Teachers can ask **Did you notice any patterns as you worked?**

| | | | | | | | | |
|---|---|---|---|---|---|---|---|---|
| **Day** | 1 | 2 | 3 | 4 | 5 | 6 | 7 | 8 |
| **Watch shows at noon** | 12:30 | 1:00 | 1:30 | 2:00 | 2:30 | 3:00 | 3:30 | 4:00 |
| **Day** | 9 | 10 | 11 | 12 | 13 | 14 | 15 | 16 |
| **Watch shows at noon** | 4:30 | 5:00 | 5:30 | 6:00 | 6:30 | 7:00 | 7:30 | 8:00 |
| **Day** | 17 | 18 | 19 | 20 | 21 | 22 | 23 | 24 |
| **Watch shows at noon** | 8:30 | 9:00 | 9:30 | 10:00 | 10:30 | 11:00 | 11:30 | 12:00 |

## Let's check!

**Main problem:** 24 days.

**Extension activity:**

48 days. The Extension activity allows for some comparison discussion around 12-hour and 24-hour recordings and why 24-hour times give more information. On a digital display, midnight is displayed as 00:00 while noon is displayed 12:00. After noon, the use of 13–23 to show hours communicates a time after noon. A display using 12-hour times does not give this information, requiring the reader of the time to understand that 12-hour times can either denote before or after noon. Some 12-hour displays may use a.m. and p.m.

## Let's reflect!

When pupils have had an opportunity to discuss their methods and solutions, teachers may wish to facilitate discussion around the way pupils approach problems. Teachers may ask questions such as **What is the difference between challenging and too challenging? Why is it important to work out an answer rather than just using a friend's answer?** These questions can offer reassurance to pupils and give teachers useful information to support pupils.

## Teacher Guide Reference

- First Level Teacher Guide Unit 7

# The Birthday Present

## Experiences and Outcomes

Number and Number Processes

- *I can use addition, subtraction, multiplication and division when solving problems, making best use of the mental strategies and written skills I have developed.* ***MNU 1-03a***

Money

- *I have investigated how different combinations of coins and notes can be used to pay for goods or be given in change.* ***MNU 1-09b***

## Resources

These should be available for pupils to use if they wish:

- play coins

## Before they start...

- Pupils should be able to add to 20.

## Let's go!

Isla is going to Nuria's birthday party. She goes to a toy shop and buys a small gift. She has no 1p or 2p coins in her purse. She hands the shopkeeper four different coins. How much might Isla have spent?

## Enabling prompt

If pupils are unable to get started on the problem, or do not understand the problem:

- Encourage pupils to use coins to explore the coins Nuria might have.
- What is the greatest and least amount that can be made using four different coins?

Encourage pupils to work as a group or with a partner to act out the situation.

## Extension activity

- How can you be certain that you have all the solutions?
- If you could use five different coins, can you predict some amounts that you could make? Try them out.

## Let's discuss!

This problem has multiple solutions and working systematically would be an advantage. Working in collaboration may support pupils in recognising the need to be systematic. It may be worth gathering the class after a short time and asking **How are you solving this problem?**

Teachers may facilitate discussion by using prompts such as:

- **How are you keeping track of your solutions?**
- **How can this help you know you have found all possible solutions?**

Some pupils may calculate that the least amount that Isla can pay is 85p, while the greatest is £3.70. They may attempt to find combinations between these two amounts. Teachers may ask **What amounts are impossible to make? How can this help you?**

Some pupils may use a table such as the one below, to systematically record their solutions. As they do so, they may notice that all amounts are multiples of 5. Teachers may ask **Why do the solutions all have 0 or 5 in the ones place?**

| 5p | 10p | 20p | 50p | £1 | £2 | |
|---|---|---|---|---|---|---|
| ■ | ■ | ■ | ■ | | | 85p |
| ■ | ■ | ■ | | ■ | | £1.35 |
| ■ | ■ | | ■ | ■ | | £1.65 |
| ■ | | ■ | ■ | ■ | | £1.75 |
| | ■ | ■ | ■ | ■ | | £1.80 |
| ■ | ■ | ■ | | | ■ | £2.35 |
| ■ | ■ | | ■ | | ■ | £2.65 |
| ■ | | ■ | ■ | | ■ | £2.75 |
| | ■ | ■ | ■ | | ■ | £2.80 |
| ■ | ■ | ■ | | ■ | ■ | £3.15 |
| ■ | | ■ | | ■ | ■ | £3.25 |
| | ■ | ■ | | ■ | ■ | £3.30 |
| ■ | | | ■ | ■ | ■ | £3.55 |
| | ■ | | ■ | ■ | ■ | £3.60 |
| | | ■ | ■ | ■ | ■ | £3.70 |

Teachers can ask **When you found a solution, how did it help you to find others? In what ways are our solutions the same?**

After the discussion, pupils may wish to gather the various solutions they have found and continue to add to them over time. There are multiple solutions to be discovered.

## Let's check!

**Main problem:** There are 15 different amounts. These are 85p, £1.35, £1.65, £1.75, £1.80, £2.35, £2.65, £2.75, £2.80, £3.15, £3.25, £3.30, £3.55, £3.60, £3.70

## Let's reflect!

When pupils have had an opportunity to discuss their methods and solutions, teachers may wish to facilitate discussion using questions such as **Did you have a plan before you started?** Some pupils may find that approaching problems in a haphazard way causes them to become demoralised. Teachers may ask **Why can it be useful to try a problem and *then* form a plan?** Teachers can encourage pupils to discuss the advantages and disadvantages of having a plan before trying a problem.

## Teacher Guide Reference

- First Level Teacher Guide Unit 3

# Ice Cream Cones

## Experiences and Outcomes

Patterns and Relationships

- I can continue and devise more involved repeating patterns or designs, using a variety of media. **MTH 1-13a**

## Resources

These should be available for pupils to use if they wish:

- pencil and paper
- 5 different colours of paper / Post-it notes / coloured multilink, etc.
- Resource sheet: Ice Cream Cones

## Before they start...

- Pupils should be able to count to 15.

## Let's go!

Isla visits the local ice cream shop. It has 5 different flavours of ice cream. How many different two-scoop ice cream cones can you make?

## Enabling prompt

- What if the shop sold only 2 ice cream flavours? How many different two-scoop cones could you buy?
- Would using coloured paper, different coloured cubes and the ice cream scoops help you 'see' the problem?
- What if the ice cream flavours are vanilla, chocolate, strawberry, mint and toffee?

## Extension activity

- What if there were 6 flavours of ice cream?
- What if there were 10 flavours?
- Can you find a pattern linking the number of flavours available and the number of scoops?

## Let's discuss!

Giving pupils time to independently puzzle over the task, try different strategies to begin the task, make errors and self-correct is important. Teachers may wish to gather pupils together *briefly* after approximately 10 minutes to share strategies and methods that are being used. Teachers may ask **What have you noticed? What have you tried? How are you recording your answers?** Allow time for pupils to share how they approached the task, what they have tried and if there were patterns that were spotted.

Some pupils may try randomly creating two-scoop ice creams with different flavours to try to work out the answer, whereas some may work systematically starting with two 'flavours' and working through all five flavours in order.

Others may choose to assign letters or numbers to the flavours, which may help them work systematically or spot a pattern. If children are struggling to write the full names of the ice cream flavours each time, they could be asked **Could we use letters to represent the ice cream flavours so we don't have to write the full word each time?** Encouraging some sort of shorthand to naming the flavours also encourages algebraic thinking.

Some pupils may not realise that a two-scoop ice cream can include the same flavour for each scoop, e.g. chocolate and chocolate. Prompts such as **Does a two-scoop ice cream have to have different flavours?** may be required for some pupils.

Problem solving can involve trialling different ideas and pupils should be allowed to find out for themselves that there is a need to record their thinking. If pupils are becoming frustrated that they cannot recall how they have solved the problem, the teacher can ask them how they can keep a record of what they did. Teachers can gather a great deal of information at this time by observing methods and conversations.

When pupils have had an opportunity to solve the problem, teachers should gather methods that pupils have used to record their thinking. Teachers may then facilitate discussion by using prompts such as:

- **What advice would you give to another class doing this problem?**
- **How did you start this problem?**
- **What helped you solve the problem (if you managed to)? What didn't help?**

## Let's check!

**Main problem:**

If we call the first flavour 1, the second flavour 2 and so on, the following answer can be derived:

| | | | | |
|---|---|---|---|---|
| 1, 1 | 1, 2 | 1, 3 | 1, 4 | 1, 5 |
| 2, 2 | 2, 3 | 2, 4 | 2, 5 | |
| 3, 3 | 3, 4 | 3, 5 | | |
| 4, 4 | 4, 5 | | | |
| 5, 5 | | | | |

| Flavour | 1 | 2 | 3 | 4 | 5 |
|---|---|---|---|---|---|
| 1 | ✓ | ✓ | ✓ | ✓ | ✓ |
| 2 | × | ✓ | ✓ | ✓ | ✓ |
| 3 | × | × | ✓ | ✓ | ✓ |
| 4 | × | × | × | ✓ | ✓ |
| 5 | × | × | × | × | ✓ |

There are 15 different two-scoop ice creams that can be bought.

**Extension activity**

| | | | | | |
|---|---|---|---|---|---|
| 1, 1 | 1, 2 | 1, 3 | 1, 4 | 1, 5 | 1.6 |
| 2, 2 | 2, 3 | 2, 4 | 2, 5 | 2.6 | |
| 3, 3 | 3, 4 | 3, 5 | 3,6 | | |
| 4, 4 | 4, 5 | 4,6 | | | |
| 5, 5 | 5,6 | | | | |
| 6,6 | | | | | |

There are 21 different two-scoop ice creams that can be bought with 6 flavours.

| | | | | | | | | | |
|---|---|---|---|---|---|---|---|---|---|
| 1, 1 | 1, 2 | 1, 3 | 1, 4 | 1, 5 | 1.6 | 1, 7 | 1, 8 | 1, 9 | 1, 10 |
| 2, 2 | 2, 3 | 2, 4 | 2, 5 | 2.6 | 2, 7 | 2, 8 | 2, 9 | 2, 10 | |
| 3, 3 | 3, 4 | 3, 5 | 3, 6 | 3, 7 | 3, 8 | 3, 9 | 3, 10 | | |
| 4, 4 | 4, 5 | 4, 6 | 4, 7 | 4, 8 | 4, 9 | 4, 10 | | | |
| 5, 5 | 5,6 | 5, 7 | 5, 8 | 5, 9 | 5, 10 | | | | |
| 6, 6 | 6, 7 | 6, 8 | 6, 9 | 6, 10 | | | | | |
| 7, 7 | 7, 8 | 7, 9 | 7, 10 | | | | | | |
| 8, 8 | 8, 9 | 8, 10 | | | | | | | |
| 9, 9 | 9, 10 | | | | | | | | |
| 10, 10 | | | | | | | | | |

There are 55 different two-scoop ice creams that can be bought with 10 flavours.

## Let's reflect!

After pupils have had the opportunity to discuss their methods and solutions, teachers may wish to facilitate discussion using questions such as **How did you feel when you first read the problem? Did your feelings change as the time progressed? What helped you persevere when things got tricky?**

## Teacher Guide Reference

- First Level Teacher Guide Unit 10

# Number Patterns

## Experiences and Outcomes

Patterns and Relationships

- Through exploring number patterns, I can recognise and continue simple number sequences and can explain the rule I have applied. **MTH 1-13b**

## Resources

- Resource sheet: Number Patterns
- Resource sheet: Number Patterns Digit Cards

## Before they start...

Pupils should be able to repeat forward number word sequences up to 40. Pupils should also have experience of skip counting starting at both odd and even numbers.

## Let's go!

Finlay has used digit cards to make a two-digit number pattern. Isla opened the window and the wind blew some of the cards and mixed them up. How might they place the digit cards to complete a pattern? How do you know?

## Enabling prompt

- Look carefully at the second number in the pattern. What can you tell me about this number? Look carefully at the fourth number in the pattern. What do you know about this number? What does that mean for the third number in the pattern?
- Try using the digit cards to create a pattern.
- What if the first number in the pattern was 25?

## Extension activity

- How might you extend Finlay's pattern?
- What if you had the cards 1, 1, 1, 1, 1, 1, 2, 3, 4, 5. What other patterns could be made?

### Let's discuss!

This is a problem that may be tackled either independently or collaboratively.

Some pupils may start by randomly placing or writing digits onto the blank cards on the Resouce sheet. This may help them to try out ideas for a pattern and discount conjectures. Asking **Can you explain this pattern?** will prompt pupils to not only check their answers but also to justify them.

Some pupils may begin by noticing that it is likely the pattern begins in the 20s and ends in the 30s. This may prompt them to use the digits in a more systematic way.

Others may use either of these strategies but with digit cards to help them visualise the number sequence.

When pupils have had an opportunity to solve the problem, teachers should gather methods that pupils have used to record their thinking. Teachers may then facilitate discussion by using prompts such as:

- **How did you start this problem?**
- **How did you ensure you were able to calculate all the answers? How did you ensure you didn't use the same card twice?**
- **How did you record your answers so that you didn't lose track?**
- **What was the first mistake you made? Did you overcome this? How?**

### Let's check!

**Main problem:** 25, 27, 29, 31, 33

**Extension activity**

Pupils should be able to justify the patterns they have created. One pattern is: 11, 12, 13, 14, 15. Another is 11, 21, 31, 41, 51.

## Let's reflect!

After pupils have had the opportunity to discuss their methods and solutions, teachers may wish to facilitate discussion using questions such as **How did you feel when you first read the problem? Did your feelings change as time progressed?** Discussing the ability to discount answers and how pupils felt when they realised their answer was not a pattern is important. The teacher may wish to ask **How did you feel when you realised you had made a mistake? How did you overcome this?** If pupils collaborated on this problem, ask **What were the advantages or disadvantages of working with others on this problem?**

## Teacher Guide Reference

- First Level Teacher Guide Unit 2

# Isla's Best Day

## Experiences and Outcomes

Expressions and Equations

- When a picture or symbol is used to replace a number in a number statement, I can find its value using my knowledge of number facts and explain my thinking to others. **MTH 1-15b**

## Resources

These should be available for pupils to use if they wish.

- blocks or other manipulatives

## Before they start...

- Pupils should be able to add within 20.

## Let's go!

The teacher decides to have a day where the children will focus on art.

'Yippee', says Finlay. 'I love art!'

'Well, I love maths, so I think this is a terrible idea!' says Isla.

'Don't worry, Isla. I have thought of that', says the teacher.

She puts this on the board.

 +  +  = ?

 +  +  = 7

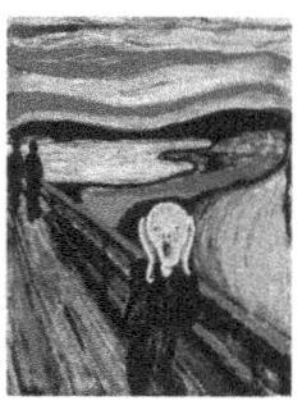 +  +  = 10

'Can you find out what numbers the ,  and  are?' says the teacher.

'Now this **is** my BEST day!' says Isla.

Can you help Isla solve the problem?

## Enabling prompt

If pupils are unable to get started on the problem, or do not understand the problem, teachers may wish to display the example below or similar and ask What does the star represent? How do you know?

★ + ★ = 20

Teachers may then look at the original problem and ask **Could the**  **be 5? Explain your thinking**. This will help teachers ensure that the children understand that the triangle represents the same number each time.

## Extension activity

- There is more than one solution, find them all.
- Create a problem like this for your partner to solve.

## Let's discuss!

For some pupils this problem can appear confusing, and it is worth looking at the enabling prompt quickly to ensure they can access the problem. When children understand that the image represents a number and the same image represents the same number, they will be able to explore possibilities.

When pupils have had an opportunity to solve the problem teachers may facilitate discussion by using prompts such as:

- **What row did you start with and why?**
- **Which piece of information was most useful?**
- **How did you know you were correct?**
- **Did you find all the possibilities? How do you know?**

Some pupils may use a 'trial and error' method. Teachers can ask **What number did you start with and why?** It is worth exploring how pupils use their initial guess to guide subsequent guesses.

Some pupils may start by looking at the last row. They may calculate all possibilities for this row before checking that their solution for a triangle and a circle are possible for the two remaining rows.

## Let's check!

**Main problem:** There are three possible solutions

 = 1  = 2  = 4 ? = 4

 = 2  = 0  = 5 ? = 4

 = 0  = 4  = 3 ? = 4

## Let's reflect!

When pupils have had an opportunity to discuss their methods and solutions, teachers may wish to facilitate discussion using questions such as **How did you use what you know about numbers to help you?** Pupils should be encouraged to recognise that they can use their existing knowledge to help them. Teachers may ask **Why is it useful to know what isn't possible?** Teachers can encourage pupils to discuss why discounting solutions early on can help them solve a problem.

## Teacher Guide Reference

- First Level Teacher Guide Unit 3

# The New Floor

## Experiences and Outcomes

Angle, Symmetry and Transformation

- I have explored symmetry in my own and the wider environment and can create and recognise symmetrical pictures, patterns and shapes. **MTH 1-19a**

## Resources

These should be available for pupils to use if they wish.

- colouring pencils
- equilateral triangle paper
- mirrors
- Resource sheet: The New Floor

## Before they start...

- Pupils should be able to identify symmetrical patterns.

## Let's go!

Nuria's dad is laying floor tiles in the kitchen. He has asked the tile shop to send him lots of red, yellow and blue triangle-shaped tiles.

He uses all three colours to create large symmetrical triangles like this:

Show some different large symmetrical triangles that Nuria's dad can make.

## Enabling prompt

- Look at the Resource Sheet: The New Floor. How do you know it is symmetrical? What happens when you turn it around?
- Use two colours.
- Use fewer triangles like this:

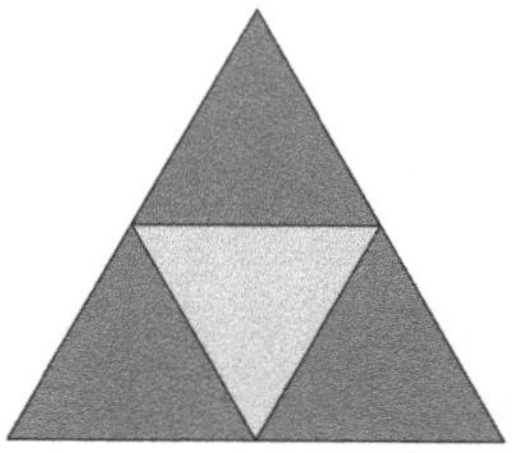

## Extension activity

- If Nuria's dad used only two colours, would there be more or fewer solutions than using three? Explain your thinking.
- Can you use three colours so that the triangle has three lines of symmetry?
- Use the equilateral triangle paper to explore other shapes that can be symmetrical. Use colour too.

## Let's discuss!

When pupils have had an opportunity to solve the problem, teachers should gather the various solutions. A visualiser may help. Teachers may wish to facilitate discussions using prompts such as

- **Are any of these solutions the same?** Teachers should ensure pupils recognise rotations of the same shape and can identify them.
- **How can we be sure the triangles are symmetrical?**
- **Can the triangle be symmetrical in different ways?**
- **What happens when I turn it round?**

Some pupils may focus on a single line of symmetry. Others may recognise that the equilateral triangle can have more than one line of symmetry.

Problem solving can be haphazard and pupils should be allowed to find out for themselves that there is a need to work systematically. After a short while, teachers may wish to gather the class and ensure that pupils are working methodically to create symmetrical patterns.

If pupils are becoming frustrated that there are many options, the teacher can ask them **What can we keep the same?** Teachers can gather a great deal of information at this time by observing methods and conversations.

## Let's check!

**Main problem:** There are multiple ways that three colours can be arranged so that there is a single line of symmetry. Teachers may wish to use a visualiser to examine some solutions in greater detail.

There is only one way that three lines of symmetry can be created using three colours.

## Let's reflect!

When pupils have had an opportunity to discuss their solutions, teachers may wish to facilitate discussion using questions such as **How did you decide what to do after you coloured the first large triangle? What helped you find different solutions?** Some pupils may find the open-ended nature of this problem a challenge as there are many possible correct solutions. Teachers may ask **What do you do when you feel you want to quit?** Teachers can encourage pupils to share strategies that they use to keep going.

## Teacher Guide Reference

- First Level Teacher Guide Unit 13

# What Colour Am I?

## Experiences and Outcomes

Properties of 2D Shapes and 3D Objects

- I have explored simple 3D objects and 2D shapes and can identify, name and describe their features using appropriate vocabulary. **MTH 1-16a**

## Resources

These should be available for pupils to use if they wish.

- pencil and paper or whiteboards for informal jottings
- Resource sheet: What Colour Am I: Blank Shapes

## Before they start...

Pupils should know the properties of circles, triangles and squares. Pupils should also be able to use the vocabulary *vertices* (corners) and *between*.

## Let's go!

What colour is each shape if:

- red has no vertices
- blue is between white and black
- blue is to the left of yellow
- green is next to white

## Enabling prompt

- Which shapes could be red?
- What if we know that one of the triangles is blue?
- What are the shape options for 'blue is between white and black'?

## Extension activity

Starting with blank shapes, how many different answers are there when the following clues are given:

- Blue is between white and black.
- Blue is on the opposite side of the shapes to green.

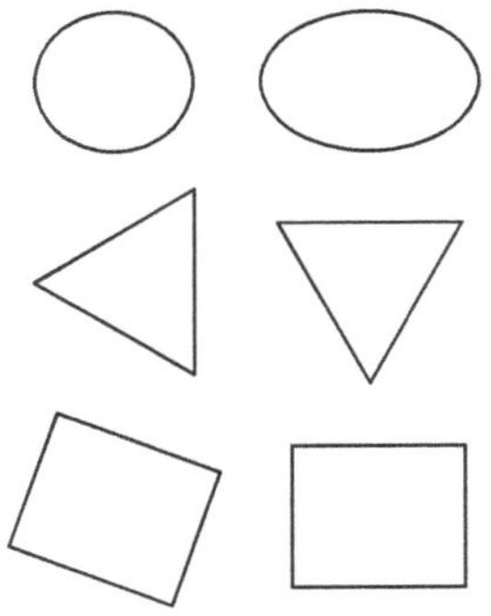

## Let's discuss!

This is a problem that may be tackled either independently or collaboratively.

Some pupils may start by drawing or colouring the shapes they are sure about (for example, blue being the left triangle and yellow being the right triangle), then filling in the rest. Some may choose to draw shapes informally then colour or use abbreviations for the colours (for example, R for red, Y for yellow). This should be encouraged as it helps promote algebraic thinking.

Others may start by taking a more random approach to the task. If pupils are colouring on paper too prematurely, they could be encouraged to use a less permanent resource; for example, a whiteboard. If pupils are struggling to show the colours of the shapes without coloured pens, the teacher may ask **How could you show that this shape is yellow without using a yellow pen? Could we write something instead?** Some may choose to write the whole colour word instead of just a letter.

For the extension activity, it may be helpful to ask **How can you make sure you have all the possible answers?** to encourage working systematically.

All pupils should be encouraged to check that their final answer matches each of the clues.

Some pupils may find it hard to persevere with this problem if they discover they have an incorrect response or have coloured a shape incorrectly early in the process. Questions such as **Where do you think you went wrong? How can we fix this? What do you need to get back on track?** may help individuals to continue with the process.

When pupils have had an opportunity to solve the problem, teachers should gather methods that pupils have used to record their thinking. Teachers may then facilitate discussion by using prompts such as:

- **How did you start this problem?**
- **What was the trickiest part of this problem?**
- **Did you get stuck? How did you get 'unstuck'?**

## Let's check!

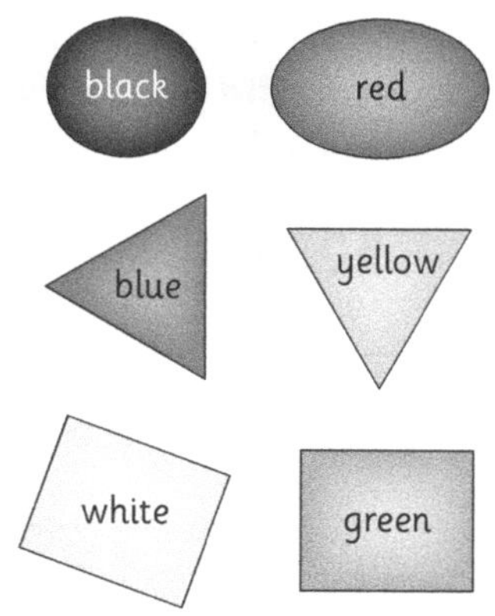


### Extension activity

There are 12 possible answers.

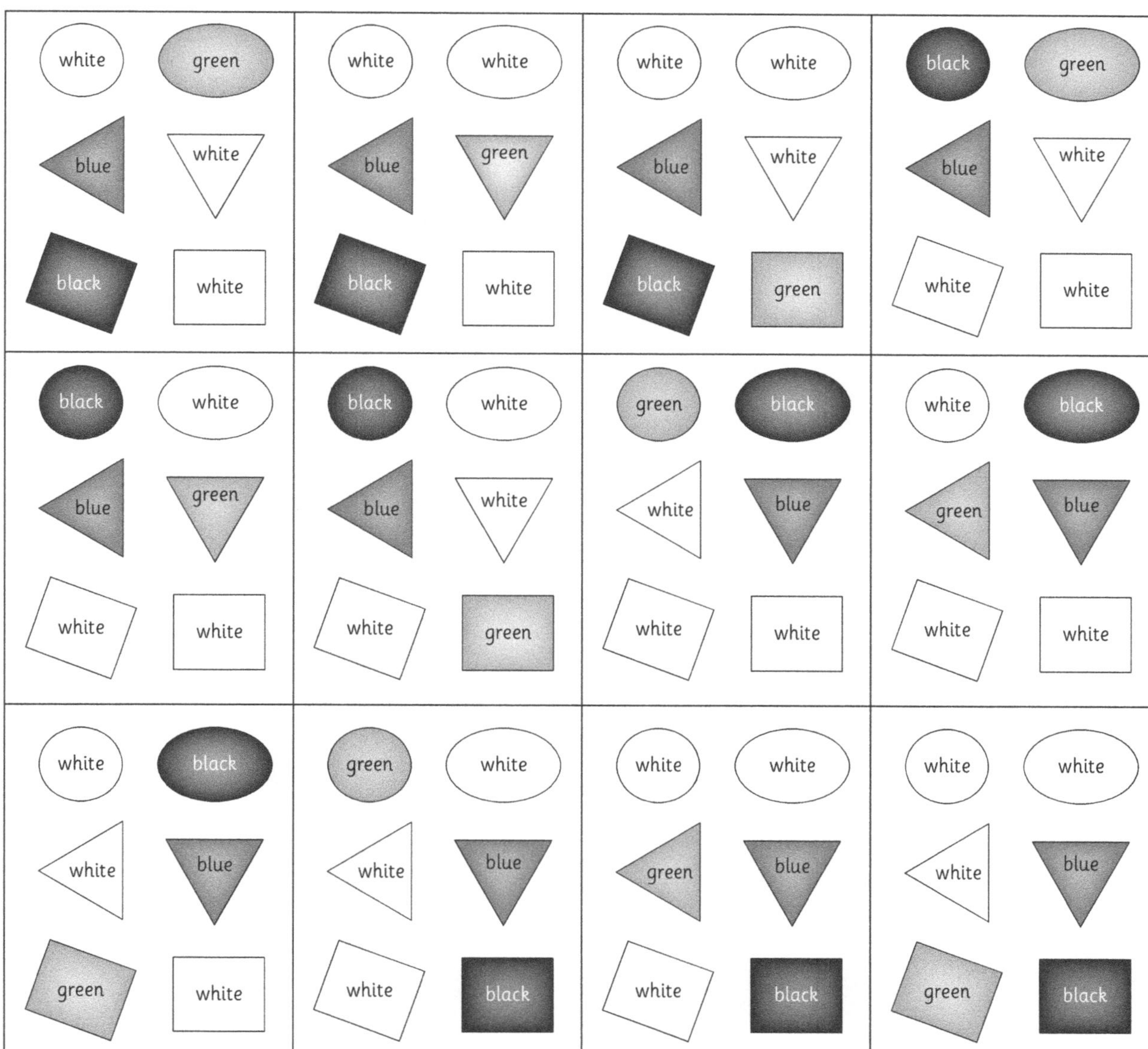

## Let's reflect!

After pupils have had the opportunity to discuss their methods and solutions, teachers may wish to facilitate discussion using questions such as **How did you feel when you first read the problem? Did your feelings change as the time progressed?** It may be useful to discuss perseverance by asking questions such as **What strategies did you use to persevere with this problem? What were the advantages and disadvantages of working with others or working on your own?**

## Teacher Guide Reference

- First Level Teacher Guide Unit 12

# Making Triangles

## Experiences and Outcomes

Properties of 2D Shapes and 3D Objects

- I have explored simple 3D objects and 2D shapes and can identify, name and describe their features using appropriate vocabulary. **MTH 1-16a**

## Resources

These should be available for pupils to use if they wish.

- strips of card or paper pre-cut to 6 cm, 5 cm, 4 cm and 3 cm lengths
- pencil and paper for jottings

## Before they start...

- Pupils need to be familiar with triangles.

## Let's go!

The children are playing in the construction area in their classroom. They see some boxes with lots of plastic strips that fit together with magnets. The plastic strips are four different lengths: 6 cm, 5 cm, 4 cm and 3 cm.

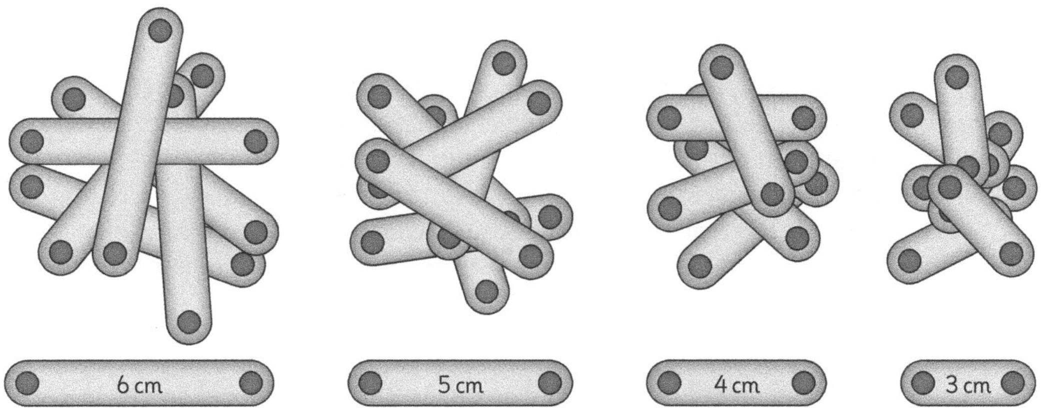

Finlay makes a triangle using two 6 cm strips and one 4 cm strip.

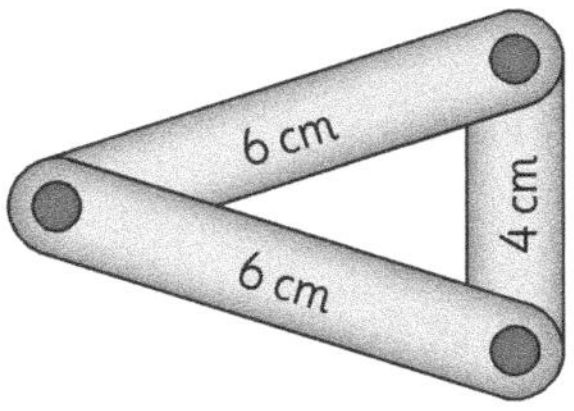


Nuria makes a different triangle. She uses one 5 cm strip, one 4 cm strip and one 3 cm strip.

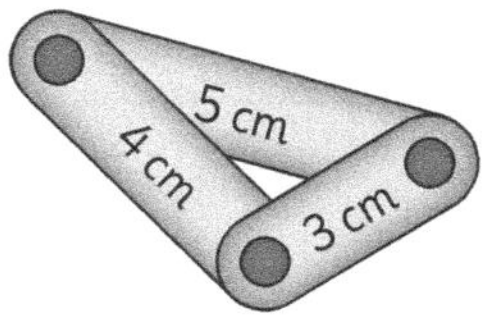


The children decide to investigate how many different triangles they can make using these plastic strips. **How many different triangles can they make?**

## Enabling prompt

Pupils who are struggling to get started can be given a simpler version of the problem involving only two different lengths of paper strip; for example, 6 cm strips and 4 cm strips, and asked to make some triangles from these.

## Extension activity

Amman chooses three strips but works out that he cannot make a triangle from them. Which strips do you think he chose?

## Let's discuss!

Some pupils are likely to begin by randomly selecting three strips, making a triangle from these then breaking this down without making a note of which strips they used. These pupils will need to be reminded of the question (**How many different triangles can the children make?**) and encouraged to make a record of each triangle. The teacher should ensure that pupils are given opportunities to share different approaches both to selecting strips and also to recording results. Where pupils are creating a table to record their results systematically, they may offer one solution more than once; for example, 6, 6, 5 and 6, 5, 6. The teacher should say **Build these two triangles for me** followed by **What do you notice?** to encourage pupils to see that the two entries are for one triangle. Allowing pupils to display different triangles on a visualiser or similar will be useful.

## Let's check!

**Main problem:**

| | | | | | | |
|---|---|---|---|---|---|---|
| 6, 6, 6 | 6, 6, 5 | 6, 6, 4 | 6, 6, 3 | 6, 5, 5 | 6, 5, 4 | 6, 5, 3 |
| 6, 4, 4 | 6, 4, 3 | 5, 5, 5 | 5, 5, 4 | 5, 5, 3 | 5, 4, 4 | 5, 4, 3 |
| 4, 4, 4 | 4, 4, 3 | 3, 3, 3 | | | | |

**Extension activity:**

One 6 cm strip and two 3 cm strips will not make a triangle.

## Let's reflect!

Pupils should reflect on how they moved on from 'building' triangles, from strips of paper or card for example, to visualising them and using a recording system to note them down. They should consider how this helped them to work efficiently. They should also think about any methods that supported them while they were working, such as 'I looked for all of the triangles with a 6 cm strip first', or 'I made equilateral triangles first', and consider why adopting a strategic approach is useful. This might then move on to consideration of the strategies they used to check that they had found all the solutions.

## Teacher Guide Reference

- First Level Teacher Guide Unit 12

# Dominoes

## Experiences and Outcomes

Number and Number Processes

- *I can use addition, subtraction, multiplication and division when solving problems, making best use of the mental strategies and written skills I have developed.* ***MNU 1-03a***

## Resources

These should be available for pupils to use if they wish.

- Resource sheet: Dominoes: Domino Square
- Real or photocopied dominoes from Resource Sheet: Dominoes from Double 3

## Before they start...

- Pupils should be able to add one-digit whole numbers within 10.

## Let's go!

Use the dominoes from double three down to double zero to make this square so that each side has eight dots.

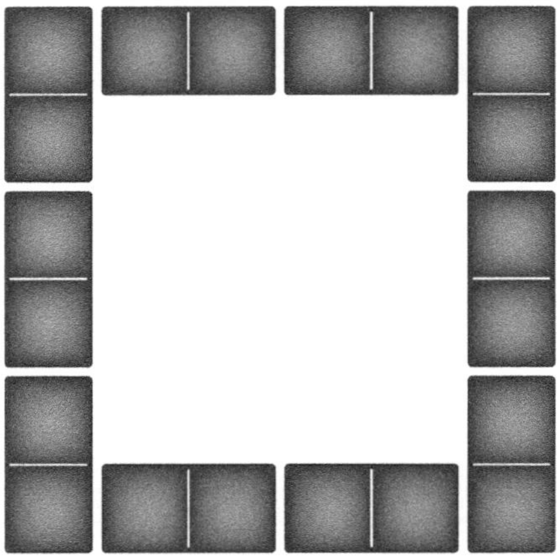

## Enabling prompt

- Which dominoes could be placed on the same side as the double three domino?

## Extension activity

- Is this the only solution you can find?
- Is there a pattern or rule for the corners of the square?

## Let's discuss!

Giving pupils time to independently puzzle over the task, try different strategies in order to begin the task, make errors and self-correct is important. Teachers may wish to gather pupils together *briefly* after approximately 10 minutes to share strategies / methods that are being used. Teachers may ask **What have you noticed? What have you tried? How are you recording your answers?** Allow time for pupils to share how they approached the task, what they have tried and if there are patterns that were spotted.

This is a problem that may be tackled either independently or collaboratively.

Some pupils may start by placing / writing dominoes randomly to make one side equal eight. The random placing of dominoes may then continue to build up the other sides.

Some pupils may use their knowledge of number properties to strategically place the dominoes. For example, they may notice that the double three has the largest total and therefore dominoes with much smaller totals must be on that side.

Problem solving can be haphazard and pupils should be allowed to find out for themselves the benefits of recording their thinking as they work through the problem. Attention can also be drawn to the advantages of having a way of recording answers that allows the pupil to know they have explored all possible options.

When pupils have had an opportunity to solve the problem, teachers should gather methods that pupils have used to record their thinking. Teachers may then facilitate discussion by using prompts such as:

- **How did you start this problem?**
- **What problems did you encounter along the way? How did you resolve them?**

## Let's check!

**Main problem:**

There are many possible solutions to this task. One solution is:

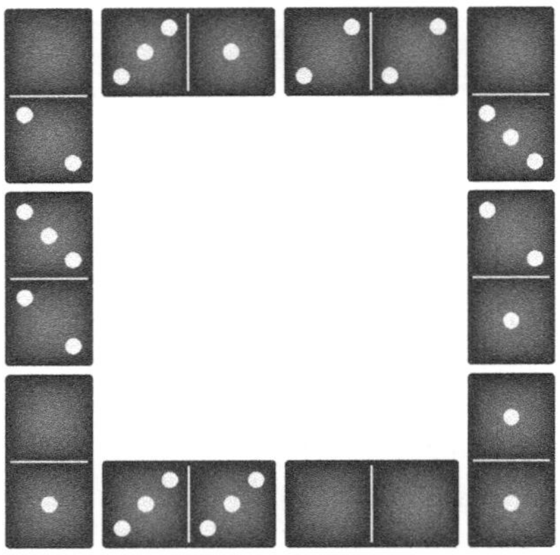

**Extension activity**

All four corners of the square must total 2, e.g. 0 + 0 + 1 + 1 as shown in the image.

## Let's reflect!

After pupils have had the opportunity to discuss their methods and solutions, teachers may wish to facilitate discussion using questions such as **How did you feel when you first read the problem? Did your feelings change as the time progressed?** If pupils collaborated on this problem, ask **What were the advantages or disadvantages of working with others on this problem?** Teachers may also wish to delve into the mistakes that pupils made and how they managed to overcome obstacles they had encountered.

## Teacher Guide Reference

- Early Level Teacher Guide Unit 3

# The New Cook

## Experiences and Outcomes

Number and Number Processes

- *I can use addition, subtraction, multiplication and division when solving problems, making best use of the mental strategies and written skills I have developed.* ***MNU 1-03a***

Data and Analysis

- *I have explored a variety of ways in which data is presented and can ask and answer questions about the information it contains.* ***MNU 1-20a***

## Resources

These should be available for pupils to use if they wish:

- blocks to represent options (yellow, white, brown and pink may help children recall which block represents which item)

## Before they start...

Pupils should be familiar with doubling numbers. Pupils should also be able to add and subtract within 20.

## Let's go!

The school cook is getting in a muddle preparing rolls for the children's class. The children have ordered ham or cheese fillings on brown or white rolls. The teacher wrote the order on this note:

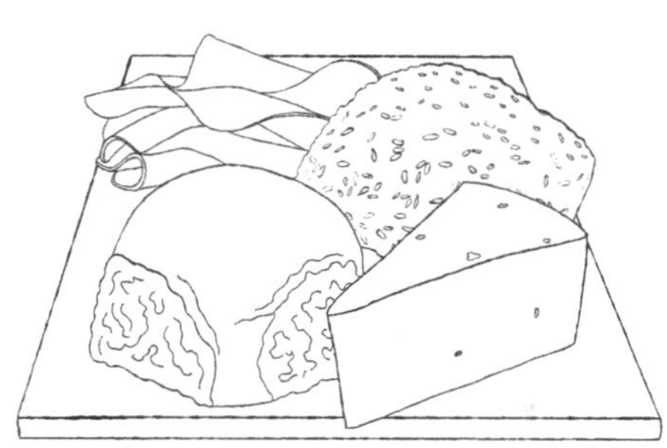

How many of each kind of roll does the cook need to make?

## Enabling prompt

If pupils are unable to get started on the problem, or do not understand the problem, one of the following prompts may help:

- What choices can there be?
- How many cheese rolls were there altogether? How many were on white rolls? How many must be on brown rolls?
- Use number sentences to help, e.g. 8 – 2 = ?
- Draw a table to help

| | **Ham** | **Cheese** | **Total** |
|---|---|---|---|
| White roll | | | |
| Brown roll | | | |
| Total | | | |

## Extension activity

- The next day, the cook offers tuna or egg rolls. There are brown and white rolls. Use the note to work out how many of each choice the cook must make.

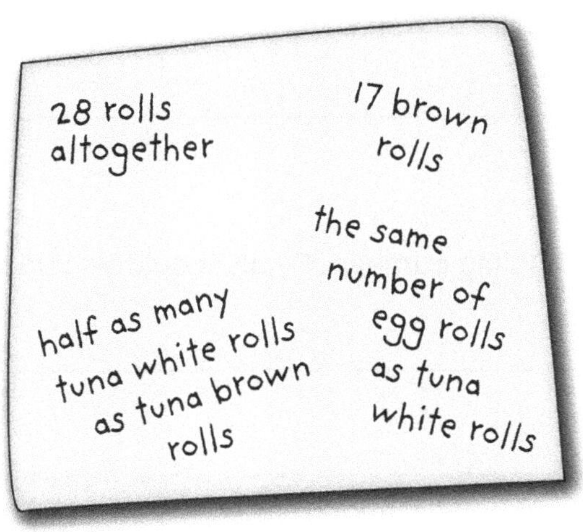


- Pupils could also be challenged to make up their own, similar problem.

## Let's discuss!

When pupils have had an opportunity to solve the problem, teachers should gather the methods that they have used to record their thinking.

Teachers can facilitate discussion by using a prompt such as:

- **What information was most useful?**
- **Did it matter which order the clues were used in?**

Pupils may use a table to organise their thinking. Teachers can draw attention to what information was given in the problem and what information had to be found. The problem asked, 'How many rolls of each kind does the cook need to prepare?'. Teachers may ask **Why is a table useful for finding *more* information?**

| | Ham | Cheese | Total |
|---|---|---|---|
| White roll | 4 | 2 | 6 |
| Brown roll | 8 | 6 | 14 |
| Total | 12 | 8 | 20 |

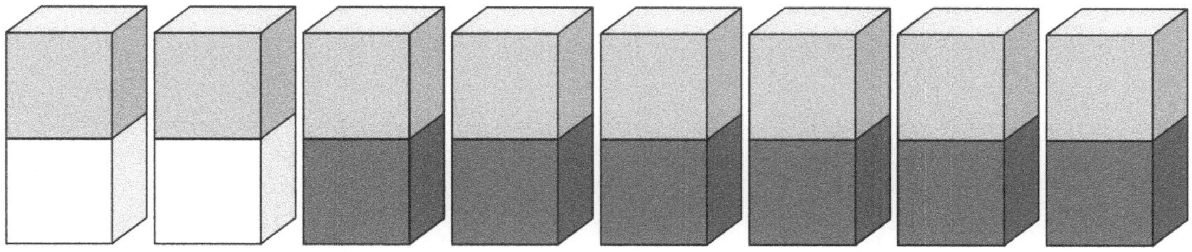

Blocks or similar can be used to represent the different ingredients. Attention should be drawn to the information contained in the problem. Teachers may ask **If we know there are only two cheese on white rolls, what does that mean for the other cheese rolls?** Pupils can discuss how they can record the information they work out.

After this discussion, children may wish to explore their strategies further using the problems generated in the extension activity above. Teachers can prompt reflection by asking **Did your original strategy work again? Have you found a different way to solve problems like these? Which is the most efficient strategy you have found?**

## Let's check!

**Main problem:**

4 white ham rolls, 8 brown ham rolls, 2 cheese white rolls, 6 brown cheese rolls.

Additional information gleaned from the table: 12 ham rolls and 8 cheese rolls. 6 white rolls and 14 brown rolls. 20 rolls in total.

**Extension activity**

14 brown tuna rolls, 7 white tuna rolls, 3 brown egg rolls and 4 white egg rolls.

## Let's reflect!

When pupils have had an opportunity to discuss their methods and solutions, teachers may wish to facilitate discussion around the information that pupils focused on in the initial stages of the problem. They may ask **How do you feel when you read a problem like this?** The acknowledgement that problems can feel overwhelming is important. Discussion can then follow as to how pupils move on from that situation.

## Teacher Guide Reference

- First Level Teacher Guide Unit 3

# Carroll Diagrams

## Experiences and Outcomes

Data and Analysis

- Using technology and other methods, I can display data simply, clearly and accurately by creating tables, charts and diagrams, using simple labelling and scale. **MTH 1-21a**

## Resources

These should be available for pupils to use if they wish.

- Resource sheet: Carroll Diagrams, Blank Carroll Diagram

## Before they start...

Pupils should be familiar with the properties of circles, triangles, rectangles and squares. They should also be able to work with whole numbers up to and including 20.

## Let's go!

The children are working together to organise these shapes into a Carroll Diagram.

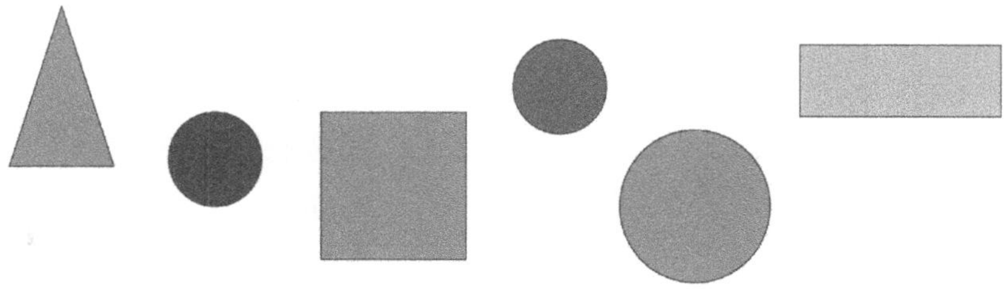

This is their completed Carroll Diagram.

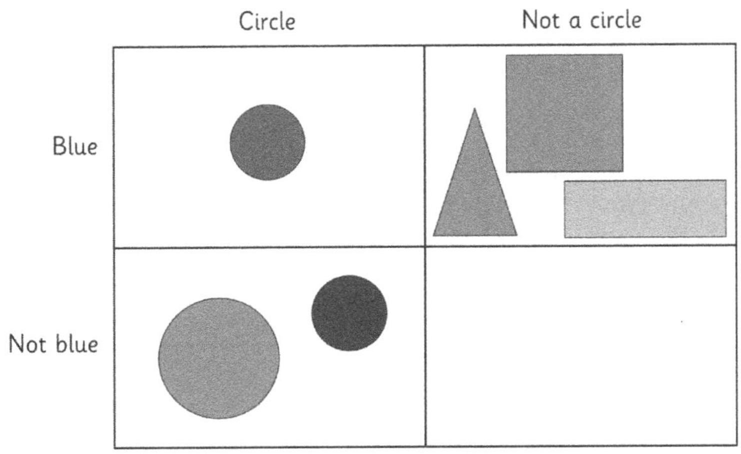


The children each decide to draw a shape that will fit into the empty cell in this Carroll Diagram.

Amman draws 

Isla draws 

Finlay draws 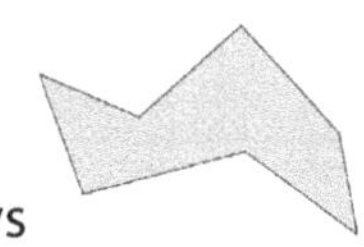

Nuria draws 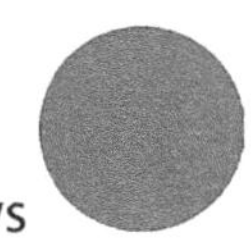

**Do each of these shapes fit into the empty cell? Explain your answers.**

Amman and Isla worked with each other to sort the numbers between 1 and 20 into a Carroll Diagram. They compared their diagram to one that Finlay and Nuria had made for the same numbers and noticed that the diagrams were different.

| | Even | Not even |
|---|---|---|
| 10 or more | 20 14 12 18 16 10 | 17 11 19 13 15 |
| Not 10 or more | 4 2 6 10 8 | 1 5 7 9 3 |

| | Even | Not even |
|---|---|---|
| 10 or more | 10 14 12 16 18 | 13 15 17 11 19 |
| Not 10 or more | 6 2 8 4 | 5 1 9 7 3 |

What do you think the children's teacher said when they checked the diagrams?

The labels on this Carroll Diagram are missing. Add labels by selecting from the different suggestions here:

Has a shell

Is not a mammal

Does not have a shell

Does not have legs

Does not live in water

Lives in water

Is a mammal

Has legs

## Enabling prompt

Pupils who are struggling to make a start to this problem may be supported by the teacher probing their understanding of the Carroll Diagram with shapes by asking questions like **Why is the purple circle here (pointing to the shape)?** The teacher may introduce one or more of the shapes below and ask **Will this fit into the empty cell? Why do you think it will / will not?**

## Extension activity

The children decide to make a Carroll Diagram to represent these foods. Create a completed Carroll Diagram to show how the children might do this.

## Let's discuss!

A brief initial discussion of the completed Carroll Diagram showing the shapes should give pupils an opportunity to revisit the Yes / No approach to sorting by two characteristics or traits. As pupils discuss the children's shapes, they should be encouraged to say why they think a shape does or does not fit. Pupils should be left to discuss the diagrams containing the numbers between 1 and 20 with little or no teacher input. The teacher should facilitate sharing of reasons why the diagrams are different, probably using the errors to highlight these. When the pupils are deciding on which labels fit into the Carroll Diagram with animals, it is likely that they will need to check some of the characteristics of the animals shown.

## Let's check!

In the first activity, Amman's shape is not a circle, but it is blue therefore it does not fit into the empty cell. Finlay's shape is not a circle and it is not blue so it fits into the empty cell. Isla's shape is not a circle and it is not blue so it fits into the empty cell. Nuria's shape is a circle so it does not fit into the empty cell.

In the activity involving the numbers between 1 and 20, each of the diagrams contains an error: the number 10 appears in two cells in the first diagram and the number 20 does not appear in the second diagram.

Adding labels activity:

The extension activity allows for various ways of labelling the Carroll Diagram, for example fruit / not fruit, vegetable / not vegetable, red / not red, orange / not orange, grows on a tree / does not grow on a tree . . .

## Let's reflect!

As pupils are working on this problem, it is likely that they will engage in ongoing discussion with each other and that this discussion will require them to check for understanding and explain their thinking. They should think about the value of collaborative thinking and discussion when working in this way. Pupils should be encouraged to reflect upon the importance of determining how reasonable their solutions were in this problem. It is likely that at times they had to think again when they realised that something did not fit in a particular place within a diagram. The need to check and be organised in their thinking should be emphasised.

## Teacher Guide Reference

- First Level Teacher Guide Unit 14